The Chemistry of the Copper and Zinc Triads

The Chemistry of the Copper and Zinc Triads

Edited by

Alan J. Welch and Stephen K. Chapman
Department of Chemistry, University of Edinburgh

ROYAL
SOCIETY OF
CHEMISTRY

The Proceedings of the First International Conference on the Chemistry of the Copper and Zinc Triads held at the University of Edinburgh, UK, 13–16 July 1992.

Special Publication No. 131

ISBN 0-85186-715-4

A catalogue record for this book is available from the British Library

Published by The Royal Society of Chemistry,
Thomas Graham House, Science Park, Cambridge
CB4 4WF

Printed in Great Britain by Bookcraft (Bath) Ltd.

Preface

This book constitutes the Proceedings of the First International Conference on the Chemistry of the Copper and Zinc Triads, held at the University of Edinburgh in July 1992.

The conference identified three main themes – Environmental Chemistry, Organometallic/Co-ordination Chemistry and Biological/Medicinal Chemistry – under which the chemistries of the six elements concerned (Cu, Ag, Au, Zn, Cd and Hg) were discussed *via* keynote lectures, invited lectures, contributed lectures and posters. Contributions from all four modes of conference presentation were solicited for this text. Although the content of the book reflects the three themes of the conference there is frequently so much natural overlap between these themes that we have chosen not to sectionalise the book.

In broad terms, therefore, the text begins with papers concerned with biological and medicinal chemistry, moves through environmmental aspects into co-ordination chemistry, then onto organometallic chemistry and cluster chemistry and finishes with a series of papers concerned with physicochemical studies of the six elements.

We believe that the fusion of several of the important aspects of the chemistries of the copper and zinc triads contributed substantially to a successful first conference and we therefore hope that this book, which mirrors the variety of chemistry discussed at the conference, will be of interest to research workers from a number of different fields.

We thank all authors for their prompt contributions, the high quality of which necessitated minimal editorial work on our part. We also thank Mrs Janet Hayes for her cheerful and patient secretarial skills.

A.J.Welch, S.K.Chapman

Contents

The Copper and Zinc Triads in Biology

R. J. P. Williams

INORGANIC CHEMISTRY LABORATORY, UNIVERSITY OF OXFORD,
OXFORD OX1 3QR, UK

1 INTRODUCTION

Of the elements with which we are concerned in this symposium, Figure 1, two have an essential significance in
biology.   They are copper and zinc. A third cadmium may
be of some importance while the other three, silver, gold
and mercury are used in chemical stains, drugs and poisons
and are not known to have any direct biological value.
They will be looked at in a separate article. Before describing the function and/or use by man or biology of the
elements it is valuable to have in mind their basic chemistry.

The special feature of the aqueous solution chemistry
of all the six elements which terminate the transition
metal series is the high affinity for inorganic sulphide or
organic thiolate. This applies no matter what the oxidation state. In the higher oxidation state these elements
may also cause ionisation of amide NH at relatively low pH.
These properties are the result of the high electron
affinity of the cations. All in all they are expected
therefore to be bound by S or N and not O donor centres.
These characteristics are maintained in biological as well
as in inorganic and analytical chemistry.

Cu   Zn
Ag   Cd
Au   Hg

Figure 1 The Copper and Zinc Triads

2 BIOLOGICAL EVOLUTION

The very high affinity of all the elements for sulphide
must have limited or even excluded their use in the
anaerobic forms of very primitive life since the atmosphere
is believed to have been rich in H_2S. In such circumstances and at pH=7 free copper ions in either oxidation state

would have been below 10^{-25}M and free zinc ions were
probably as low as 10^{-15}M, Figure 2. The availability of
other metal ions which have more soluble sulphides, such
as Mn^{2+}, Fe^{2+} and even Co^{2+} and Ni^{2+} would have been
considerably greater. The simplest inspection of living
forms most closely related to early anaerobic life,
prokaryotes and archaebacteria, indicate that they contain
less copper and zinc than advanced eukaryote and multicellular organisms. The suggestion is strong that copper
and zinc concentrations changed radically with the advent
of dioxygen and the consequent removal of H_2S some 1-2
billion years ago. The change in the composition of the
air introduced many changes in the chemical composition of
the waters in which life developed. It is also the case
that as well as the oxidation of copper and zinc sulphides
there would have been oxidation of CdS to give some cadmium in waters. There has never been much exposure of
life to silver, gold or mercury except for minor incidents of poisoning. They are of low abundance and of high
insolubility with many anions.

     It would appear that we have to see the major evolu-
tion of biological chemistry some 1-3 billion years ago in
terms of a switch from sulphide to oxide solubility following the introduction of dioxygen most probably due to a
biological activity itself, Figure 3. In other words some
species created a massive pollution of the original sulphide dominated surface of the earth, killing off many
other species and forcing others into the present-day
anaerobic sulphide rich slime at the bottom of lakes and
the sea. This massive pollution was accompanied by a
dramatic increase in availability of some elements, including Cu and Zn, all of which would initially be as
poisonous as Ag, Au, Cd and Hg are today. The change was
also accompanied by a massive drop in the availability of
iron, precipitated by the presence of dioxygen as ferric
hydroxide, from the pre-existing ferrous sulphide solution,
and considerable changes in the species of chemically
available sources of elements like S, Se and Mo. Evolution
is partly based on random DNA mutation but this has to be
seen as seeking advantage from environmental conditions and
their changes. Evolution follows environmental change even
though it can help to bring it about. New chemicals such
as dioxygen, copper or zinc ions follow a progression of
biological activity following their gradual introduction.

     POISON $\longrightarrow$ PROTECTIVE REACTION $\longrightarrow$ FUNCTIONAL USE

To appreciate the glory of the biological chemistry of
copper and zinc today we must remember the chain of
events, the history, of incorporation of these "poisons".

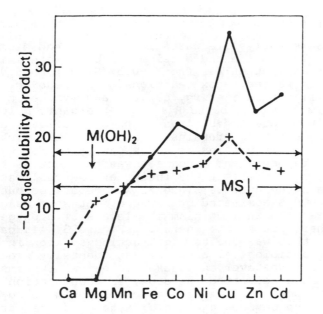

Figure 2 The solubility products of divalent sulphides and hydroxides. The horizontal lines give the products at which precipitation occurs at pH = 7 and for 10^{-3}M metal ions, see reference.

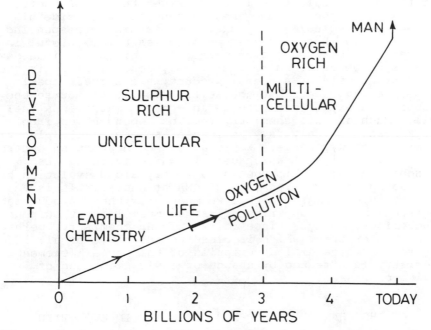

Figure 3 A schematic illustration of the evolution of life, development, with time.

3 EARLY LIFE

I shall assume early life was sulphide rich and dioxygen
very poor. The inside and outside of cells were maintained at redox potentials from -0.5 (H_2/H^+) to 0.0
(H_2S/S) by energised redox reactions. The redox catalysis essential for this energy capture and synthesis
($H_2S \longrightarrow S_n$ + bound H; bound H + $CO_2 \longrightarrow$ organic polymers)
was probably maintained by iron/sulphide catalysts as we
see today in the Fe/S proteins. Quite possibly nickel and
cobalt had ancillary roles as redox catalysts for hydrogen
and carbon fragments and manganese was the best weak Lewis
acid catalyst. In effect the inside of the cell was more
reducing than the outside since S_n was deposited outside
cells though H_2S dominated everywhere. At some stage, but
through some unknown mechanism the levels of dioxygen in
the atmosphere started to increase, Figure 3. It can be
argued that this was due to light acting on iron in the
sea or that a biological catalyst, accidentally produced,
was involved. Whatever the cause biological systems soon
developed and accelerated the dioxygen accumulation. At
first the response of adaption by cells would be to generate protective devices against dioxygen and its partially
reduced products based on existing iron and manganese
sources. Since O_2 easily goes to superoxide and hydrogen
peroxide in the presence of metal ions such as Fe and Mn,
and these chemicals are more dangerous than O_2 itself, we
expect protection to develop first against them and then
for these chemicals as well as dioxygen to become useful.
We know that iron and manganese dependent superoxide dismutases evolved early in prokaryotes as did corresponding
catalases. We know too that peroxidases became useful in
the generation of secondary metabolites giving various new
poisons of other organisms, e.g. polyphenols. This is the
first production of a useful product, a chemical capable
of destroying others but not self. Finally of course the
excess of dioxygen was used as an energy source in a fuel
cell. Much of this chemistry remains dependent on iron
and manganese catalysis. So long as some species (plants)
generated dioxygen there would always be developing others
(plants and animals) which used it while avoiding its
poisonous nature. Dioxygen also vastly aided evolution by
increasing the mutation rate of DNA of course. It is
against this background that we must view the introduction
of free zinc and copper but first we must look at sulphur
biochemistry changes since it is sulphur chemistry, perhaps
with that of iron, which is close to the heart of all life.
Figure 4 is a general description of the way all element
chemistry is affected by change of environment and of
living processes.

4 THE CHANGES IN SULPHUR CHEMISTRY IN EVOLUTION

Sulphide dominated early life in many ways and today's life
is still dominated inside cells by the reactions of sulphur

at low potential. The cycles coupled to these reactions
are largely of carbon and hydrogen at potentials between
0.0 and -0.5 volts. While the major carriers of the
hydrogen redox reactions are pyridines, quinones, and
flavins which frequently connect to Fe/S proteins or even
Fe haem proteins in the cell, the transfer of carbon fragments and energy may well have rested with thiolate esters.
It is the case today that the sulphur chemistry inside
cells remains thiolate or thioester rather than RSSR
chemistry. It is also the case that iron chemistry in
cells remains as it was originally to a large degree.
The introduction of dioxygen meant there was and is a
constant fight against its damage to this in-cell machinery. Outside the cells the coming of dioxygen raised the
effective redox potential slowly toward +0.8v. The chemistry of sulphur then became that of -S-S- and, as in
the present-day sea, of SO_4^{2-}. The value of the -S-S-
link outside the cell is seen in the hundreds of extracellular proteins stabilised by -S-S- cross-links much
as they are used to harden rubber for tyres. The only
way RS^-, thiolate could then be stable outside cells
became via a sufficiently strong complexing agent. This
is where the chemistry of copper (and molybdenum) becomes
so fascinating, but note the changes in many non-metals
as well as of metals, Table 1.

        5   THE BIOCHEMISTRY OF COPPER

Immediate inspection of the biochemistry of copper shows
that it occurs almost invariably outside cells. By outside the cell we must include not just strictly extracellular space but the inside of reticula, vesicles, and
periplasmic space, Figure 5, where the redox potential may
be thought of as approaching +0.8v rather than below 0.0
volts. Here it is that copper becomes so valuable since
the redox potential of the Cu^+/Cu^{2+} system is almost
invariably above +0.2 volts. The reason for this high
potential is that Cu^+ binds organic molecules through S
or N with equal affinity to that of the binding of Cu^{2+}.
Copper could only be useful in redox reactions of life
after dioxygen was available because of this feature of
its chemistry. It did not become useful in the cytoplasm since all such potentials are avoided. Iron by
way of contrast as Fe^{2+}/Fe^{3+} is extremely useful in the
cytoplasm at potentials below +0.0 volts since Fe^{3+} is
a much stronger Lewis acid than Fe^{2+}. Bound iron became
useful at high redox potentials only through its unstable FeO complexes. The coming of dioxygen plus the
coming of copper gave evolution an enormous chance which
it did not fail to find - it discovered extracellular
high potential redox catalysis in the form of bound copper
held by RS^- or nitrogen donor ligands. This chemistry is
often free-radical redox chemistry since outside cells
such chemistry is not a danger to DNA (mutations) while
most importantly it allows free-radical polymerisation

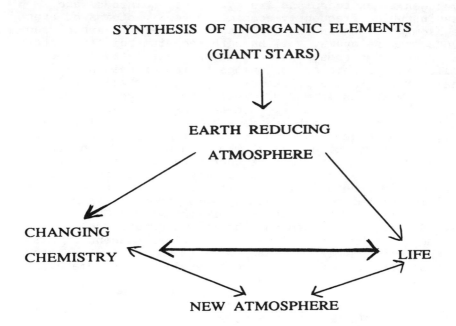

<u>Figure 4</u> A general scheme for the evolution of the
properties of elements on the earth.

<u>Table 1</u>   Non-Metals After the Advent of Dioxygen

O	$\longrightarrow$	O_2, O_3
CO, CH_4	$\longrightarrow$	CO_2
NH_3	$\longrightarrow$	N_2 (N/O)
H_2S	$\longrightarrow$	S_n, SO_4^{2-}
H_2Se	$\longrightarrow$	SeO_4^{2-}
$(MoS_4^{2-}$	$\longrightarrow$	$MoO_4^{2-})$
(Halides	$\longrightarrow$	Halogens)

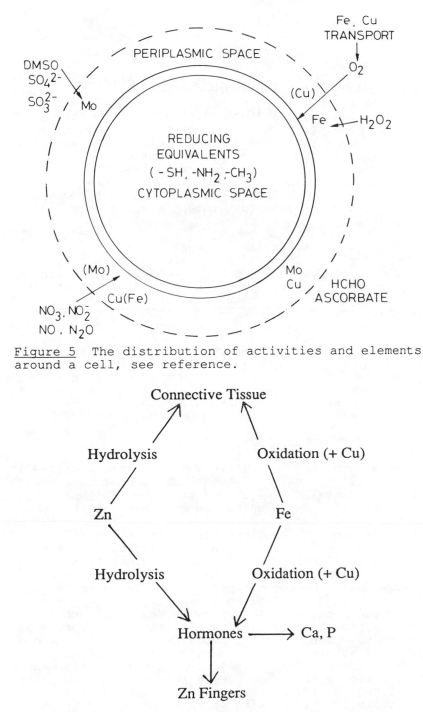

<u>Figure 5</u> The distribution of activities and elements around a cell, see reference.

<u>Figure 6</u> The connections between zinc and copper and both hormones and connective tissue. Note the involvement of iron and calcium too.

and cross-linking of polymers. Cu/O_2 chemistry vastly
increased biological polymer chemistry moving it from
the condensation polymerisation inside cells (flexible
polymers such as DNA, RNA, proteins and polysaccha-
rides) to free radical induced cross-linked polymerisation
outside cells giving connective tissue (chitin, collagen,
elastin etc.) Thus through Cu/O_2 biology evolved multicellular organisms, see Figure 6. I have only time to
mention in passing the other values of Cu/O_2 reactions in
energy capture (cytochrome oxidase), electron transfer in
vesicles (thylakoids and periplasmic space), in the
metabolism of oxides of nitrogen (periplasmic space), in
the production of secondary metabolites (adrenalin in
vesicles), and in the transport of dioxygen (haemocyanin).
All are related to high redox potential chemistry. All
are effectively extracellular.

     There is one exceptional use of copper where it acts
inside the cells. This is in the superoxide dismutase of
eukaryotes. The switch from Fe/Mn to Cu/Zn superoxide
dismutases is a very tell-tale event in evolution since
the only good reasons for such a switch are (1) A change
in availability of Cu and Zn with a lowering of that of
Fe (2) A required lowering of mutation risk in the switch
to a very strongly held divalent (and monovalent) metal
ion, copper, from a more weakly bound divalent redox
metal, Mn or Fe. Both the latter carry a high mutation
risk. It is very revealing that this switch is not seen
in what are thought to be primitive systems where mutation
could be more helpful than harmful.

     6   ZINC IN BIOCHEMISTRY

Zinc could have been incorporated to some small degree in
very primitive biological systems but it must have been
difficult to obtain. The coming of dioxygen changed that.
We know that zinc found quite new uses but unlike copper
it is used both inside and outside cells. It is not a
redox threat to DNA. Again it found binding sites in both
S and N coordination spheres where competition from Fe or
Mn is slight. The increase in zinc concentration allowed
it, in the absence of copper which was banished to outside
the cell, to become the strongest intra-cellular Lewis
acid. (Note the use of nickel chemistry lessened since
the relative insolubilities of sulphides Zn>Ni but oxides
Ni>Zn). Noticeably zinc is found in many S/N centres in
the cells of organisms developed after the coming of dioxygen i.e. eukaryotes and multicellular organisms but not
so much in prokaryotes. Parallel with this development in
evolution zinc became a very powerful part of regulatory
genes at the transcription level, zinc fingers and similar
proteins, whereas iron switched to translational control
from its role as transcriptional control in prokaryotes.
Zinc also became a stabilising cross-linker for intracellular proteins, e.g. transcarbamylase.

The strength of zinc as a Lewis acid has made it
extremely valuable as a catalytic element, Figure 6
as well as in structural/regulatory functions. Here the
role is mainly in hydrolytic reactions of peptide and
ester bonds but it is also important in RNA synthesis and
reverse transcriptase, i.e. in synthetic pathways, and in
redox two-electron reactions of some NADH dependent enzymes. Quite striking is the role in the generation from
precursor proteins of hormonal peptides (ACTH and encepha-
lins for example), their destruction by extracellular
enzymes and in the degradation of connective tissue polymers, e.g. collagen. Clearly these features associated
with signalling and growth of multi-cellular systems came
after the advent of dioxygen. The connection with the
breakdown of connective tissue in extra-cellular enzymes,
stabilised by -S-S- links also come after dioxygen upsurge,
leads immediately to thoughts about copper chemistry and
its complementary role in the stabilisation of connective
tissue by oxidative cross-linking. It is tempting to con-
clude that the roles of sulphur, copper and zinc evolved
to give a system for the mend (Cu) and cut (Zn) operations
within connective tissue necessary for the controlled
building of relatively rigid multicellular systems. The
combination of the two "new" metals with a new role for
sulphur is made the more intriguing by the link in the
handling of the homeostasis of copper and zinc via the
intracellular thiolate-rich protein, metallothionein.
While copper is handled as Cu(I) and extruded from the
cell, zinc is retained as Zn(II), and cadmium is simply
sequestered. It may well be that through metallothionein
there is a joint homeostatic balance of both zinc and
copper in which lies the secret of connective tissue and
the evolution of multicellular organisms.

Since copper and zinc evolved with sulphide chemistry
it is of interest to enquire further how they can work
together rather than compete for ligands.

7 ZINC AND COPPER COMPLEX ION CHEMISTRY

The differential binding of copper and zinc cannot be
related to a single factor. Undoubtedly copper can be
selected as the only monovalent ion, Cu^+, of high Lewis
acid strength which has become available to biology. In
vitro Ag^+ can be made into a competitor. This oxidation
state is associated with a cation of a large radius, 1.0Å,
when Cu^+ is easily separated from Zn^{2+} (0.65Å) through the
use of a reagent with a prescribed hole size in a protein
structure. This trick is used in analytical chemistry too
and is parallel to the distinction made between Ca^{2+} (1.0Å)
and Mg^{2+} (0.65%) binding to proteins.

However Cu^{2+} is also a better Lewis acid than Zn^{2+}
(Irving-Williams series). It is easily the most stable
cation with ligands which are multidentate N-donors, and

especially when they give rise to N-based anions as in superoxide dismutase and albumins. Clearly zinc can best find unique sites if copper is first removed locally to very low concentrations. In cells free copper Cu^{2+} is below $10^{15}M$, reduced and removed as Cu^+ in metallothionein, and it is here that Zn^{2+} dominates since as a free ion it is at $10^{-10}M$. Biological complex ion chemistry uses all the subtle methods man has found in his analytical studies.

8 HORMONES AND ZINC

The division of hormonal chemistry in relationship to these two metal ions is also revealing. Zinc is associated with the synthesis and degradation of peptide hormones but also with the hormone receptors which bind to DNA. The hormones concerned are the sterols, retinoic acids, thyroxines etc. Most of these hormones are secondary metabolites produced by iron enzymes in cells. The systems are absent from prokaryotes. It appears as if direct control of growth processes via free iron speaking to DNA in prokaryotes has been replaced by hormonal control via hormones produced by iron, which speak to DNA via zinc receptors. While iron may help the homeostasis of prokaryotes directly it does so indirectly in a multicellular organism via zinc. The zinc itself speaks to connective tissue and peptide hormones and also speaks to copper.

Copper has a minor role in hormonal chemistry being linked to adrenaline hydroxylation. Here the hormone (transmitter) is produced and stored in vesicles but note that a prior step of production is connected to iron.

9 THE INTERACTIVE ROLE OF ELEMENTS IN CELLS

While this artice has concentrated on copper and zinc and their interactions and there are clear links to sulphur and iron chemistry it would be a msitake to see these systems in isolation. The cell is in a homeostatic condition which links many elements together, Figure 6. A way of seeing this is to follow the connection from either iron or zinc to calcium which is another element associated with connective tissue and supporting mineral structure. The connection from calcium is via the vitamin D hormones produced by iron enzymes and interacting with zinc finger receptors. From calcium there is an immediate link to phosphate (and magnesium) which involves also zinc, magnesium and calcium dependent phosphatases. In order to appreciate this biochemistry fully it will be necessary to trace the source of proteins back to the DNA regulation. It is here that the functions of individual elements meet in the controls over the expression of coded information.

10 SUMMARY

Copper and zinc may not have been much involved with the origin of life which may well be related to iron and sulphur functions. These two elements are intimately involved in the major step to organised multicellular life however when the role of oxidation changed. The full ramifications of these discoveries are as yet unknown.

REFERENCES

This lecture is an extension of considerations considered and analysed in the book "The Biological Chemistry of The Elements" by J.R.R. Frausto da Silva and R.J.P. Williams, Oxford University Press, Oxford (1991).

Electronic Structures of Active Sites in Copper Proteins: I. Blue Copper Sites

Edward I. Solomon* and Michael D. Lowery

DEPARTMENT OF CHEMISTRY, STANFORD UNIVERSITY, STANFORD, CA 94305, USA

1 INTRODUCTION

It has now been ten years since a general review of our Group's research in the field of copper proteins has been prepared.[1] Over this period, our understanding of the electronic structure of the active sites in these proteins has strongly evolved, and is providing significant insight into the reactivity of these sites in biology. The first goal of my sabbatical was to generate an overview of our present understanding of this field, which has now appeared in *Chemical Reviews*[2] and will be summarized in two parts.[3] Part I of this summary is being published as our contribution to the Proceedings of the International Conference on the Chemistry of the Copper and Zinc Triads, held in Edinburgh, Scotland July 13-16, 1992. Part II is submitted for the Proceedings of the Symposium on Copper Coordination Chemistry: Bioinorganic Perspectives, August 3-7, 1992.

Many of the most important classes of active sites in copper proteins exhibit unique spectral features compared to simple high-symmetry transition metal complexes. These derive from the unusual geometric and electronic structures which can be imposed on the metal ion in a protein site. It has been the general goal of our research to understand these electronic structures and evaluate their contributions to the reactivities of these active sites in catalysis.

Our contributions to four topics will be summarized. First, if one is to understand the origin of unique spectral features, one must first understand the electronic structure of normal high symmetry transition metal complexes. For us, square planar cupric chloride has served as an electronic structural model complex. It is now one of the most well-understood molecules in Inorganic chemistry[4], and its spectral features and associated electronic structure will be briefly described in Part I.[3a] Then the unique spectral features of the blue copper active site will be addressed and used to provide insight into ground and excited state contributions to long-range electron transfer in these proteins. In Part II[3b] of this summary we focus on the coupled binuclear copper proteins, hemocyanin and tyrosinase, which have similar active sites that generate the same oxy intermediate involving peroxide bound to two copper(II)'s. The hemocyanins reversibly bind dioxygen while the tyrosinases have highly accessible active sites which bind phenolic substrates and oxygenate these to *ortho*-diphenols. Their oxy sites exhibit unique excited state spectral features which reflect new peroxide copper-bonding interactions which make a significant electronic contribution to the binding and activation of dioxygen by these sites. Finally, the multicopper oxidases laccase, ascorbate oxidase and ceruloplasmin catalyze the four electron reduction of O_2 to water. Spectral studies on these enzymes will also be summarized in Part II which demonstrate that they contain a fundamentally different coupled binuclear copper site (called Type 3) which, in fact, is part of a trinuclear copper cluster which plays the key role in the multielectron reduction of dioxygen by this important class of enzymes.

2 NORMAL COPPER COMPLEXES

Starting with normal copper complexes [Figure 1], placing a cupric ion with its nine d electrons in an octahedral ligand field produces a ^2E_g ground state which is unstable to a Jahn-Teller distortion which lowers the symmetry and energy of the complex. The Jahn-Teller distortion normally observed is a tetragonal elongation along the z axis and contraction in the equatorial x,y plane ultimately resulting in a square planar ligand environment as in D_{4h}-$CuCl_4^{2-}$ [Figure 1A]. A key feature of ligand field theory[5] is that the d orbital splitting is very sensitive to the environment of the ligands around the metal center. The d orbital splitting experimentally observed for D_{4h}-$CuCl_4^{2-}$ from optical spectroscopy[6] is given in Figure 1B where the highest-energy half-occupied orbital is $d_{x^2-y^2}$ as it has the largest repulsive interaction with the ligands in the equatorial plane. A more complete description of this half-occupied ground state is provided by molecular orbital (MO) theory; in particular self consistent field-$X\alpha$-scattered wave (SCF-$X\alpha$-SW) calculations[7] adjusted to ground state parameters[8-11] as will be described, provide good agreement with spectral data over many orders of magnitude in energy. These calculations[8,9] generate a description of the ground state of D_{4h}-$CuCl_4^{2-}$ which has 61% Cu $d_{x^2-y^2}$ character with the remaining part of the wavefunction being delocalized equivalently into the four p_σ orbitals of the chloride ligands which are involved in antibonding interactions with the metal ion [Figure 1C]. The unpaired electron in this wavefunction produces the EPR spectrum shown to the right in Figure 1D where $g_{\parallel}$ (corresponding to the magnetic field oriented along the z axis of the complex) $> g_{\perp} > 2.00$ is characteristic of this $d_{x^2-y^2}$ ground state. Additionally, copper has a nuclear spin of 3/2 which couples to the electron spin to produce a four line hyperfine splitting of the EPR spectrum. Tetragonal cupric complexes generally have a large $A_{\parallel}$ value; that of D_{4h}-$CuCl_4^{2-}$ is 164×10^{-4} cm^{-1}.[12]

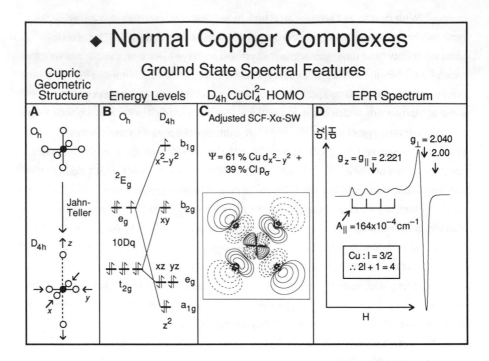

Figure 1 Normal Copper Complexes: (A) Jahn-Teller tetragonal elongation of an octahedral CuL_6 complex to D_{4h} symmetry. (B) Energy level correlation diagram for the Jahn-Teller distortion depicted in A. (C) SCF-Xα-SW wavefunction contour and charge decomposition for the HOMO of D_{4h}-$CuCl_4^{2-}$. (D) X-band EPR spectrum for tetragonal Cu(II) with D_{4h}-$CuCl_4^{2-}$ parameters.

With respect to excited states, optical excitation of electrons from the filled d orbitals to the half-filled $d_{x^2-y^2}$ orbital produces ligand field transitions[13] which, as shown in Figure 2A, are Laporté forbidden and hence weak in the absorption spectrum with ε's of 30-50 M^{-1} cm^{-1} in the 12,000-16,000 cm^{-1} region[6] [Figure 2B, right]. To higher energy in the absorption spectrum are the Laporté allowed ligand-to-metal charge transfer transitions which are at least two orders of magnitude more intense than the ligand field transitions [Figure 2B, left]. The energies and intensities of these charge transfer transitions allow one to probe the specific bonding interactions of the ligand with the metal center.[14] Chloride has three valence 3p orbitals which split into two sets on binding to the copper [Figure 2C]. The p_σ orbital is oriented along the Cl-Cu bond and is stabilized to deep binding energy due to strong overlap. The two p_π chloride orbitals are perpendicular to the Cl-Cu bond, hence more weakly interacting with the metal and at lower binding energy. The intensity associated with charge transfer excitation of an electron from these filled ligand orbitals into the half-occupied Cu $d_{x^2-y^2}$ orbital also reflects bonding and is proportional to $(RS)^2$, where S is the overlap of the donor and acceptor orbitals involved in the charge transfer transition. Thus, the Cl $p_\sigma \rightarrow$ Cu $d_{x^2-y^2}$ charge transfer transition is at high energy and intense, while the Cl $p_\pi \rightarrow d_{x^2-y^2}$ charge transfer is weak and at lower energy due to poor overlap [Figure 2C]. The key points to be emphasized here are that the charge transfer transitions sensitively probe the ligand-metal bond, and that for normal complexes one should observe a lower energy weak π and higher energy intense σ charge transfer transition.

3 BLUE COPPER PROTEINS

As anticipated by spectroscopy[15], the blue copper site has a structure very different from the normal tetragonal geometry of cupric complexes. The copper site in plastocyanin[16] has a distorted tetrahedral structure with thiolate sulfur of Cys 84 bound with a short Cu-S bond length of 2.13Å, thioether sulfur of Met 92 bound with a long Cu-S bond length of 2.90Å and two fairly normal histidine N-Cu ligands [Figure 3A]. It is well known that this site has quite unique spectral features[1,2] including an intense absorption band of $\varepsilon \sim 3,000$-5,000 M^{-1} cm^{-1} in the 600 nm ligand field region [Figure 3B] and a small parallel hyperfine splitting of $\leq 70 \times 10^{-4} cm^{-1}$ [Figure 3C]. These unusual spectral features are now well understood and help to define the ground state wave-function of the blue copper site. This is extremely important in that this is the half-occupied orbital which takes up and transfers an electron in the redox functioning of this center. A detailed experimental and theoretical description of the ground state thus provides fundamental insight into the active site contribution to long-range electron transfer by the blue copper proteins.[2]

Since the EPR spectrum in Figure 3C shows $g_\parallel > g_\perp > 2.00$, the blue copper site must have a $d_{x^2-y^2}$ ground state and we are first interested in defining its orientation

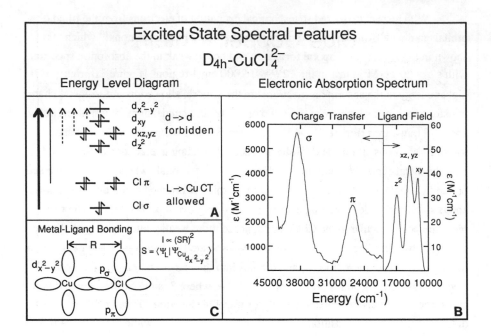

Figure 2   Excited state spectral features of D_{4h}-$CuCl_4^{2-}$: (A) Energy level diagram showing the ligand field (d → d) and charge transfer optical transitions. The intensity of the transitions is approximated by the thickness of the arrow with the very weak ligand field transitions represented as a dotted line. (B) Schematic of the σ and π bonding modes between the Cu $3d_{x2-y2}$ and Cl 3p orbitals. (C) Electronic absorption spectrum for D_{4h}-$CuCl_4^{2-}$ (adapted from references 6 and 14).

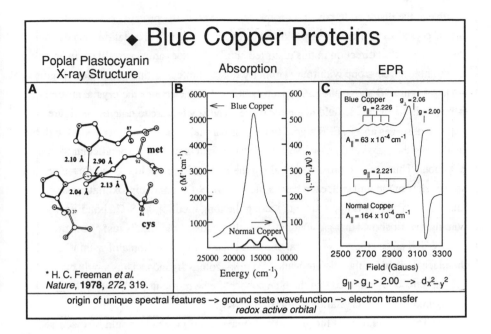

Figure 3   Blue Copper proteins: (A) X-ray structure of poplar plastocyanin. (B) Absorption spectrum of plastocyanin and "normal" D_{4h}-$CuCl_4^{2-}$ (ε scale expanded by 10). (C) X-band EPR spectrum of plastocyanin and D_{4h}-$CuCl_4^{2-}$.

relative to the distorted tetrahedral geometry observed in the protein crystal structure. Single crystal EPR spectroscopy allowed us to determine this orientation thus defining the unique (i.e. z) direction in this distorted site.[17] Plastocyanin crystallizes in an orthorhombic space group with four symmetry related molecules in the unit cell [Figure 4A].[16] Figure 4B shows the EPR spectra obtained if one rotates the crystal around the a axis with the magnetic field in the b/c plane. The key feature to note in the figure is that one observes an $\sim g_{\parallel}$ EPR signal (with four parallel hyperfine components) with the field along the c axis and an $\sim g_{\perp}$ spectrum as the field is rotated perpendicular to this direction. Thus, $g_{\parallel}$ is approximately along the c axis. Referencing to the four blue copper sites in the unit cell, each has its methionine S-Cu bond approximately along this c axis. Thus, $g_{\parallel}$ is approximately along the long methionine S-Cu bond. A more quantitative fit shows that $g_{\parallel}$, which defines the z axis of the site, is just 5° off this bond which places the $d_{x^2-y^2}$ orbital perpendicular to this direction and within 15° of the plane defined by the thiolate S and two imidazole N ligands. Thus single crystal EPR spectroscopy required that the blue copper site be reoriented relative to that initially configured by X-ray crystallography [Figure 4C].

The next feature of the ground state wavefunction which should be discussed is the origin of small $A_{\parallel}$ hyperfine splitting. Often distorted tetrahedral cupric sites, for example D_{2d}-CuCl$_4^{2-}$, exhibit small $A_{\parallel}$ values similar to the blue copper proteins and these have been thought to have a common origin. In D_{2d}-CuCl$_4^{2-}$, the small $A_{\parallel}$ has been attributed to the effect of Cu 4p mixing into the $d_{x^2-y^2}$ orbital which is allowed in lower symmetry metal sites.[18] The idea here is that in D_{2d} symmetry, the $4p_z$ orbital can mix into the $d_{x^2-y^2}$ orbital by Group Theory and its spin dipolar interaction with the copper nuclear spin will oppose that of the electron spin in the $d_{x^2-y^2}$ orbital, thus lowering the $A_{\parallel}$ value. 12 % $4p_z$ mixing is required to lower $A_{\parallel}$ to the value observed for D_{2d}-CuCl$_4^{2-}$ [19] and plastocyanin.[20] Thus we are interested in determining the 4p mixing into the $d_{x^2-y^2}$ of plastocyanin. A combination of ligand field theory[17] and low temperature MCD spectroscopy[24] has given the splitting of the d orbitals at the blue copper site which probes the effective symmetry of its ligand field. Five nondegenerate levels are observed indicating a rhombically distorted site, however the splitting is close to an axial limit. The splitting in Figure 5 corresponds to a C_{3v} axially elongated tetrahedral structure where the z axis corresponds to the long thioether S-Cu bond. However, in C_{3v} symmetry, the $d_{x^2-y^2}$ orbital is only allowed to mix with the $4p_x$, $4p_y$ levels and in this case the spin dipolar interactions would be complementary and hence increase the $A_{\parallel}$ value.

Thus we needed to directly determine the nature of the 4p mixing into the $d_{x^2-y^2}$ level. This was accomplished by going up about ten orders of magnitude in photon energy and performing X-ray absorption spectroscopy at the Cu-K edge.[21] The idea here is that the 8979eV pre-edge peak corresponds to the Cu 1s → $3d_{x^2-y^2}$ transition and can

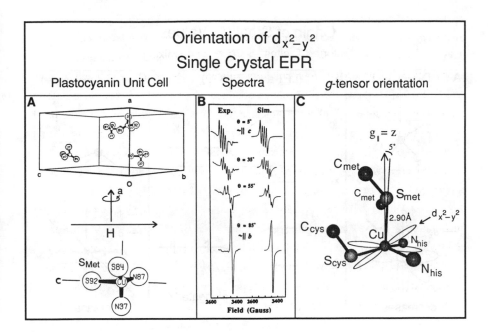

Figure 4   Single Crystal EPR of poplar plastocyanin[17]: (A) Unit cell and molecular orientation with respect to the applied magnetic field. (B) EPR spectra and simulations for the crystal orientations shown. (C) Orientation of the $g_{\parallel}$ axis and $d_{x^2-y^2}$ orbital superimposed on the blue copper site.

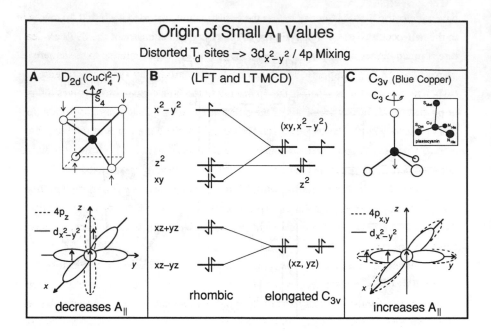

<u>Figure 5</u>   Origin of small $A_\parallel$ values: (A) $4p_z$ mixing with the d_{x2-y2} orbital in D_{2d} symmetry lowers $A_\parallel$ (B) Energy levels for blue copper site (left) and its axial C_{3v} limit (right). (C) Effect of $4p_{x,y}$ mixing with the d_{x2-y2} orbital on $A_\parallel$ for a C_{3v} distortion.

have no electric dipole intensity unless the copper site is distorted such that 4p mixes into the half-occupied $d_{x^2-y^2}$ level [Figure 6A]. Indeed the intensity of the 8979 eV feature is much higher in the plastocyanin and D_{2d}-$CuCl_4^{2-}$ edges relative to the square planar D_{4h}-$CuCl_4^{2-}$ complex[21] which can have no 4p mixing by symmetry [Figure 6B]. One can determine whether the 4p mixing in the blue copper site involves the p_z or p_x, p_y orbitals through analysis of the polarized single crystal X-ray absorption spectra of plastocyanin. The single crystal EPR data[17] defined the z axis as the long thioether S-Cu bond and polarized edge data were obtained[22] with the E vector of light parallel and perpendicular to this direction. No 8979 eV intensity is observed for E ∥ z requiring that there is no $4p_z$ mixing, while all the 8979 eV intensity is observed in the E ∥ *x,y* spectrum. Thus only Cu $4p_x$,p_y mixes into the $d_{x^2-y^2}$ orbital of plastocyanin and the small $A_\parallel$ value of the blue copper site cannot be due to the generally accepted mechanism of $4p_z$ mixing.

Having eliminated $4p_z$ mixing, we can focus on the alternative explanation for the small $A_\parallel$ value of the blue copper site which is covalent delocalization of the electron spin onto the ligands;[23,24] this reduces its hyperfine interaction with the copper nuclear spin. We have approached the inclusion of covalency in the description of the ground state of the blue copper site through a quantitative consideration of its g values. Multi-frequency EPR spectroscopy[23] gives the experimental g values for plastocyanin listed on the left in Table I. If the ground state only involved an unpaired electron in a $d_{x^2-y^2}$ orbital, there would only be a spin angular momentum contribution to the g values and these would be 2.00 and isotropic. Ligand field theory[5] allows for the inclusion of some orbital angular momentum into the $d_{x^2-y^2}$ ground state through its spin-orbit mixing with ligand field excited states. This orbital contribution to the g values leads to the deviations from 2.00. A complete ligand field calculation for plastocyanin gives the g values listed in the third column of Table I with $g_\parallel > g_\perp > 2.00$ consistent with a spin orbit mixed $d_{x^2-y^2}$ ground state. However, the calculated values are quantitatively larger than the experimental values. This is due to the fact that the ligand field calculations use pure d orbitals which have too much orbital angular momentum. Covalent delocalization of the unpaired electron onto the ligands reduces its orbital contribution to the g values.

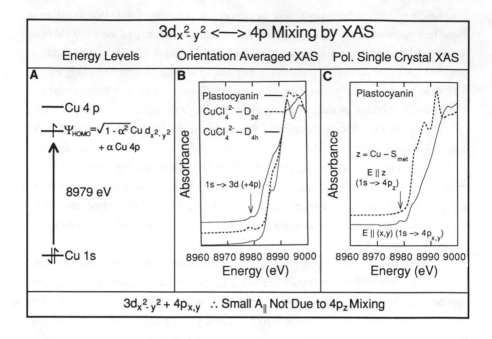

Figure 6   X-ray Absorption Spectroscopy: (A) Energy level diagram depicting a Cu $1s \rightarrow$ HOMO transition at ~8979 eV. (B) Orientation averaged XAS spectra. (C) Polarized single crystal XAS spectra for poplar plastocyanin.[22]

Table I. Blue Copper Covalency –> Quantitative Analysis of g Values

	experimental	$(x^2-y^2)^1$ spin only	L. F. T. d orbitals $+ \lambda \, L \cdot S$	SCF-Xα-SW d levels, CT levels $\lambda_{Cu} \, L \cdot S + \lambda_L \, L \cdot S$ Norman radii	Adjusted radii
g_x	2.047	2.00	2.125	2.046	2.059
g_y	2.059	2.00	2.196	2.067	2.076
g_z	2.226	2.00	2.479	2.159	2.226

Thus SCF-Xα-SW calculations[23,24] were pursued to describe the bonding in the blue copper site. The wavefunctions obtained from these calculations were then used to calculate the ground state g values and thus evaluate the covalent description generated by these calculations relative to experiment. The g value calculation included all the antibonding (d) and bonding (charge transfer) levels and involved spin-orbit mixing from both the metal and the ligands. A Zeeman operator was then applied to the spin-orbit corrected ground state making no assumption concerning the orientation of the principal axes. A g^2 tensor was generated and diagonalized to obtain the principal component g values which can then be compared to experiment (fourth column). It is observed that while the g values are indeed reduced from those obtained from the ligand field calculation due to covalency, they are in fact closer to 2.00 than is obtained experimentally. Thus these SCF-Xα-SW calculations are producing too covalent a description of the active site. There is one set of adjustable parameters in this calculation which is the sphere sizes used in the scattered wave solutions. Those employed in this initial calculation are the standard spheres normally used which are defined by their Norman radii.[25] We then systematically varied these spheres (increasing the metal sphere which increases its electron density, lowers its effective nuclear charge and reduces its interaction with the ligands, while matching potentials at the sphere boundaries) and iteratively repeated this g value protocol until the calculated values were in reasonable agreement with experiment.[23,24] This approach provided the experimentally adjusted description of the ground state wavefunction of the blue copper site which is given in Figure 7A.

These Xα calculations provided a description of the ground state of the blue copper site which is highly covalent and the covalency is highly anisotropic involving the Sp$_\pi$ orbital of the thiolate [Figure 7A]. We have now been able to experimentally test these key features of this ground state using a variety of spectroscopies. First, the high covalency can be probed by copper L-edge spectroscopy.[26] The idea here is that

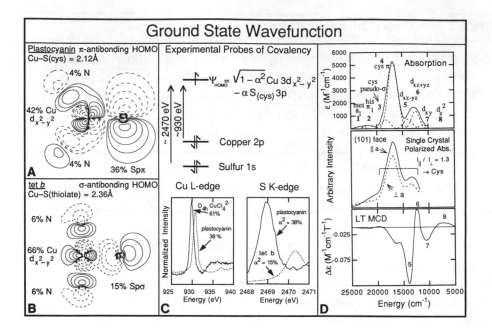

Figure 7   Ground state wavefunction of plastocyanin: (A) HOMO wavefunction contour for plastocyanin. (B) HOMO wavefunction contour for the thiolate copper complex tet *b*.[27] (C) Copper L-edge and sulfur K-edge spectra as a tool to experimentally measure the metal-ligand covalency. (D) Absorption, single crystal polarized absorption and low-temperature MCD spectra of plastocyanin. The absorption spectrum has been Gaussian resolved into its component bands as in reference 24.

the electric dipole intensity of the Cu 2p → Ψ_{HOMO} transition at 930 eV reflects the Cu 2p → Cu 3d transition probability and thus probes the amount of Cu $d_{x^2-y^2}$ character in the ground state wavefunction. From Figure 7C it is observed that the 930 eV peak in plastocyanin has about 2/3 of the intensity of D_{4h}-CuCl$_4^{2-}$, which from the beginning section of this presentation has 61% Cu $d_{x^2-y^2}$ character. Thus the blue copper site is estimated from experiment to have 38% Cu $d_{x^2-y^2}$ character in agreement with the adjusted SCF-Xα-SW calculations. The sulfur contribution to the HOMO can be studied using sulfur K-edge spectroscopy[21] where now the electric dipole intensity reflects the S1s → S3p character in the HOMO. From Figure 7C, plastocyanin exhibits an intense sulfur pre-edge feature at 2469 eV. It has an ~2 1/2 times the intensity of the tet *b* model complex of Prof. H. Schugar[27] which contains a normal 2.36 Å copper-thiolate sulfur bond and has ~15% Sp character in the ground state. Thus the blue copper site is also experimentally calibrated to have 38% sulfur p character from the cysteine ligand again in good agreement with the Xα calculations.

The final feature of the ground state wavefunction is elucidated through the assignment of the unique excited state absorption spectral features of plastocyanin. While there are in fact eight bands required to fit a combination of absorption, CD and MCD spectra of the blue copper site[24], at low resolution the absorption spectrum was originally regarded as having a low-energy weak and higher-energy intense (i.e., 600 nm) band pattern.[1,15] Polarized single crystal spectral studies over this region[17] showed the same polarization ratio for both bands, the value of which required that both bands be associated with the cysteine thiolate-copper bond. Thus, in parallel to the Cl → Cu charge transfer assignment presented earlier, these were assigned as low-energy weak π and higher energy intense σ charge transfer transitions involving the thiolate sulfur. However, low-temperature MCD spectroscopy showed that all four of the low energy bands (5-8) which comprise this region are weak in the absorption spectrum but quite intense in the MCD spectrum.[24] Since this C-term intensity for copper(II) requires spin-orbit coupling, hence d character, this leads to the assignment of bands 5-8 as d → d transitions. Thus the 600 nm band 4, which is intense in the absorption spectrum and weak in the low-temperature MCD spectrum, is the lowest energy charge transfer transition from the thiolate and must be the cys p_π → Cu $d_{x^2-y^2}$ charge transfer transition, with the Cys p_σ → Cu $d_{x^2-y^2}$ transition being to higher energy and weak. The key point here is that for the blue copper site one has a low-energy intense π and higher energy weak σ charge transfer transition to the Cu $d_{x^2-y^2}$ orbital, and since charge transfer intensity reflects orbital overlap this requires that the $d_{x^2-y^2}$ orbital have its lobes bisected by the Cu-S$_{cys}$ bond [Figure 7A] and thus be involved in a strong π antibonding interaction with the thiolate as also obtained from the Xα calculations. The strong π interaction rotates the $d_{x^2-y^2}$ orbital by 45° relative to its usual orientation along the

ligand-copper bond as for example in the tet *b* model complex [Figure 7B] and derives
from the quite short blue copper Cu-S_{cys} bond length of 2.13 Å.

Thus the SCF-Xα-SW calculations are producing a quite good description of
the ground state of the blue copper site and one can now correlate this with crystal
structure information to obtain significant insight into function. In particular, Messerschmidt, et al.[28] have now published the X-ray structure of ascorbate oxidase which
shows that the Cys S ligand of a blue copper site in this multicopper oxidase is flanked
on either side in the sequence by histidines which are ligands to two of the coppers in a
trinuclear copper cluster site which will be discussed in Part II.[3b] This blue copper site
transfers an electron rapidly in the reduction of O_2 at the trinuclear copper center. As
can be seen from the Xα calculated wavefunction contours which we have superimposed on the crystal structure of the blue center in ascorbate oxidase, the ground state
wavefunction provides a highly anisotropic covalent pathway involving the cysteine
sulfur. In addition, the low energy intense cys π→ Cu $d_{x^2-y^2}$ charge transfer transition
in the blue copper absorption spectrum provides an efficient hole superexchange pathway for rapid electron transfer between the blue and trinuclear copper cluster sites.[2]
Clearly from Figure 8 the unique electronic structure of the blue copper center reflects a
ground state wavefunction which plays a critical role in its functioning of rapid long-
range electron transfer to a specific location in or on the protein.

4 COUPLED BINUCLEAR COPPER PROTEINS AND MULTICOPPER OXIDASES: SEE PART II[3b]

5 SUMMARY

At this point the unique spectral features associated with the major classes of active sites
in copper proteins are reasonably well understood and define active site electronic
structures which provide significant insight into their reactivities in biology. For the
blue copper sites we have determined that the unique spectral features derive from a
ground state wavefunction which has a high anisotropic covalency involving the thiolate ligand. This provides a very efficient superexchange pathway for long range electron transfer. For the coupled binuclear copper active sites we have seen that the unique
spectral features of the oxy site correspond to a new bridging peroxide electronic structure which has very strong σ donor and π acceptor properties. These appear to make
significant contributions to the reversible binding and activation of dioxygen by these
active sites. In the multicopper oxidases, our spectral studies have determined that the
Type 3 center is fundamentally different from the coupled binuclear copper site in
hemocyanin and tyrosinase, that is it is in fact part of a trinuclear copper cluster and that
this trinuclear copper cluster is the structural unit required for O_2 reactivity. We have
now defined a peroxide level intermediate at this trinuclear copper cluster site which is

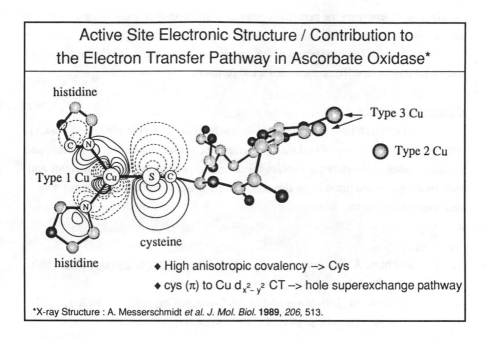

Active Site Electronic Structure / Contribution to the Electron Transfer Pathway in Ascorbate Oxidase*

histidine

Type 3 Cu

Type 2 Cu

Type 1 Cu

cysteine

histidine

♦ High anisotropic covalency –> Cys

♦ cys (π) to Cu $d_{x^2-y^2}$ CT –> hole superexchange pathway

*X-ray Structure : A. Messerschmidt *et al. J. Mol. Biol.* **1989**, *206*, 513.

Figure 8 Proposed electron transfer pathway in blue copper proteins. The plasto-cyanin wavefunction contours have been superimposed on the blue copper (Type 1) site in ascorbate oxidase.[28] The contour shows the substantial electron delocalization onto the cysteine Sp_π orbital which activates electron transfer to the trinuclear copper clu-ster at 13Å from the blue copper site. This low energy, intense Cys $S_\pi \rightarrow$ Cu charge transfer transition provides an effective hole superexchange mechanism for rapid long-range electron transfer between these sites.[2]

strikingly different from the peroxide bound in oxyhemocyanin and oxytyrosinase. Our spectral studies presently underway should provide important insight into the geometric and electronic structure differences which are reflected by these spectral differences and their contribution to differences in biological function.

Acknowledgments

This research has been supported by the NSF (CHE-8919867) for the blue copper studies and by the NIH (DK-31450) for the coupled binuclear and multicopper oxidase studies. EIS wishes to express his sincere appreciation to all his students and collaborators who are listed as co-authors in the literature cited for their commitment and contributions to this science.

References

1. E. I. Solomon, K. W. Penfield and D. E. Wilcox, <u>Structure and Bonding</u>, 1983, <u>53</u>, 1.

2. E. I. Solomon, M. J. Baldwin and M. D. Lowery, <u>Chem. Rev.</u>, 1992, <u>92</u>, 521.

3. (a) This work. (b) to be published in the Proceedings of the Symposium on Copper Coordination Chemistry: Bioinorganic Perspectives, August 3-7, 1992.

4. E. I. Solomon, <u>Comments on Inorg. Chem.</u>, 1984, <u>3</u>, 227.

5. General references for ligand field theory are:
    (a) C. J. Ballhausen, C. J. "Introduction to Ligand Field Theory", McGraw-Hill, New York, 1962. (b) D. S. McClure, "Electronic Spectra of Molecules and Ions in Crystals", Academic Press, New York, 1959. (c) J. S. Griffith, "The Theory of Transition Metal Ions", Cambridge University Press, London, 1964. (d) S. Sugano, Y. Tanabe and H. Kamimura "Multiplets of Transition Metal Ions in Crystals", Academic Press, New York, 1970. (e) B. N. Figgis, "Introduction to Ligand Fields", Interscience, New York, 1967.

6. M. A. Hitchman and P. J. Cassidy, <u>Inorg. Chem.</u>, 1979, <u>18</u>, 1745.

7. General references for SCF-Xα-SW molecular orbital theory are:
    (a) K. H. Johnson, <u>Advan. Quantum Chem.</u>, 1973, <u>7</u>, 143. (b) K. H. Johnson, J. G. Norman, Jr. and J. W. D. Connolly, "Computational Methods for Large Molecules and Localized States in Solids", F. Herman, A. D. McLean and R. K. Nesbet, Eds., Plenum, New York, 1973. (c) J. W. D. Connolly, "Semiempirical Methods of Electronic Structure Calculation, Part A: Techniques", G. A. Segal, Ed., Plenum, New York, 1977. (d) N. Rosch, "Electrons in Finite and Infinite Structures", P. Phariseu and L. Scheire, Eds., Plenum, New York, 1977. (e) J. C. Slater, "The Calculation of Molecular Orbitals", Wiley, New York, 1979.

8. A. A. Gewirth, S. L. Cohen, H. J. Schugar and E. I. Solomon, <u>Inorg. Chem.</u>, 1987, <u>26</u>, 1133.

9. E. I. Solomon, A. A. Gewirth and S. L. Cohen, "Understanding Molecular Properties", J. Avery, *et al.* Eds., D. Reidel, Dordrecht, 1987; pp 27.

10. S. V. Didziulis, S. L. Cohen, A. A. Gewirth and E. I. Solomon, <u>J. Am. Chem. Soc.</u>, 1988, <u>110</u>, 250.

11. A. Bencini and D. Gatteschi, <u>J. Am. Chem. Soc.</u> 1983, <u>105</u>, 5535.

12. C. Chow, K. Chang and R. D. Willett, <u>J. Chem. Phys.</u>, 1973, <u>59</u>, 2629.

13. E. I. Solomon, M. D. Lowery, L. B. LaCroix and D. E. Root, accepted for publication in <u>Methods in Enzymology</u>.

14. S. R. Desjardins, K. W. Penfield, S. L. Cohen, R. L. Musselman and E. I. Solomon, <u>J. Am. Chem. Soc.</u>, 1983, <u>105</u>, 4590.

15. E. I. Solomon, J. W. Hare and H. B. Gray, <u>Proc. Natl. Acad. Sci. U.S.A.</u>, 1976, <u>73</u>, 1389.

16. J. M. Guss and H. C. Freeman, <u>J. Mol. Biol.</u>, 1983, <u>169</u>, 521.

17. K. W. Penfield, R. R. Gay, R. S. Himmelwright, N. C. Eickman, V. A. Norris, H. C. Freeman and E. I. Solomon, <u>J. Am. Chem. Soc.</u>, 1981, <u>103</u>, 4382.

18. C. A. Bates, W. S. Moore, K. J. Standley and K. W. H. Stevens, <u>Proc. Phys. Soc.</u>, 1962, <u>79</u>, 73.

19. M. Sharnoff, <u>J. Chem. Phys.</u>, 1964, <u>41</u>, 2003.

18. C. A. Bates, W. S. Moore, K. J. Standley and K. W. H. Stevens, <u>Proc. Phys. Soc.</u>, 1962, <u>79</u>, 73.

20. J. E. Roberts, T. G. Brown, B. M. Hoffman and J. Peisach, <u>J. Am. Chem. Soc.</u>, 1980, <u>102</u>, 825.

21. S. E. Shadle, J. E. Penner-Hahn, H. J. Schugar, B. Hedman, K. O. Hodgson and E. I. Solomon, submitted to <u>J. Am. Chem. Soc.</u>

22. R. A. Scott, J. E. Hahn, S. Doniach, H. C. Freeman and K. O. Hodgson, <u>J. Am. Chem. Soc.</u>, 1982, <u>104</u>, 5364.

23. K. W. Penfield, A. A. Gewirth and E. I. Solomon, <u>J. Am. Chem. Soc.</u>, 1985, <u>107</u>, 4519.

24. A. A. Gewirth and E. I. Solomon, <u>J. Am. Chem. Soc.</u>, 1988, <u>110</u>, 3811.

25. J. G. Norman, Jr. <u>Mol. Phys.</u>, 1976, <u>31</u>, 1191.

26. S. J. George, J. Chen, M. D. Lowery, E. I. Solomon and S. P. Cramer, submitted to <u>J. Am. Chem. Soc.</u>

27. J. L. Hughey IV, T. G. Fawcett, S. M. Rudich, R. A. Lalancette, J. A. Potenza and H. J. Schugar, <u>J. Am. Chem. Soc.</u>, 1979, <u>101</u>, 2617.

28. A. Messerschmidt, R. Ladenstein, R. Huber, M. Bolognesi, L. Avigliano, R. Petruzzelli, A. Rossi and A. Finazzi-Agró, <u>J. Mol. Biol.</u>, 1992, <u>224</u>, 179.

Structure and Activity of Type 1 Cu Sites

G. W. Canters, A. P. Kalverda, and C. W. G. Hoitink

GORLAEUS LABORATORIES, LEIDEN UNIVERSITY, PO BOX 9502,
2300 RA LEIDEN, THE NETHERLANDS

1. Introduction

It is customary to distinguish Cu sites in Cu containing proteins into different types (type 1, 2 or 3) with the implicit assumption that sites of one type are similar. In this contribution this assumption is considered for type 1 sites by looking at azurins and amicyanins, blue copper proteins that have been under investigation in the metallo-protein group in Leiden. It appears that there is more diversity among type 1 sites than a simple taxonomic categorisation would suggest, and that amicyanins have more in common with another category of type 1 blue copper proteins, plastocyanins, than with azurins. Azurins and amicyanins show clear mutual differences in mechanistic behaviour. This contribution will be used to track these differences and to see if they correlate with differences in structure of the Cu sites. A related question that is interesting, but difficult to answer, is what inferences the available structural and mechanistic information allows with regard to the evolutionary connections between the three classes of proteins mentioned above.

2. 3D structures

Azurins (Azu's) occur in aerobic as well as in anaerobic redox chains of a variety of purple bacteria. Their position in the redox chain is near the end close to the terminal electron acceptor. The 3D-structures of the Azu's from *Pseudomonas aeruginosa* (P.a.) and *Alcaligenes denitrificans* (A.d.), as well as those of a number of mutants, have been solved by X-ray diffraction techniques.[1-7] Azu's consist of a polypeptide chain of 128-129 amino acids which is folded into two β-sheets. Together, they form a β-sandwich. An α-helix comprising approximately the residues 53-67 is the only other secondary structure element present. Complete ^{1}H and ^{15}N NMR assignments have been achieved for the P.a and A.d. Azu's in solution.[8,9] The secondary structures as derived from the NMR analyses are in virtually complete accordance with the X-ray data.[8]

Plastocyanins (Pc's) are found in cyanobacteria and in the chloroplasts of plants and green algae where they mediate et between the cyt b_6f complex and the P700 complex of the photosynthetic reaction centre. The structure of Pc from *Populus nigra* has been reported in great detail by Freeman and co-workers.[10,11] NMR studies by Wright and coworkers[12] confirm the similarity between solution structures and X-ray structures. The protein roughly shows the same overall β-sandwich structure as Azu but the loop that contains the α-helix in Azu is much less pronounced and shorter in Pc.

Amicyanins (Amc's) have been isolated from methylotrophic bacteria, where they occur in electron transfer (et) paths that connect dehydrogenases with terminal electron acceptors. Their length has been found to vary from 99 to 106 amino acids.

NMR studies have shown that the secondary structure of Amc has more in common with those of Pc's than of Azu's.[13,14,14a] Preliminary results of complete tertiary structure elucidations by X-ray and NMR techniques confirm this conclusion.[14b,15,16]

When the polypeptide chains of the three proteins are aligned the Cu site is found roughly in the same excentric position at the so-called "Northern end" of the protein, about 5-7 Å beneath the protein surface (see figure 1).

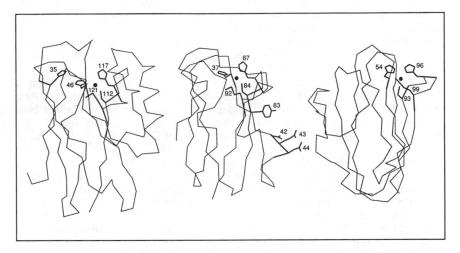

Figure 1. Backbone tracings of (from left to right) Azu (P.a.)[4,6], Pc (*P. nigra*)[10] and Amc (*T. versutus*).[15] The Cu ligands His-46, Cys-112, His-117 and Met-121 in Azu have been labeled together with residue His-35. In Pc the Cu-ligands His-37, Cys-84, His-87 and Met-92, and the residues Asp-42, Glu-43, Asp-44 and Tyr-83 in the acidic patch have been marked. The preliminary Amc structure is based on NMR data and is the result of an distance geometry calculation with 642 distance constraints, 99 angle constraints and 16 stereospecific assignments. Cu ligands His-54, Cys-93, His-96 and Met-99 have been marked. The Cu atom in the structures is represented by a black circle.

3. Electron transfer; paths and activity

The function of blue copper proteins is to transfer electrons. The metal site in its oxidised form is able to accept an electron and to subsequently transfer it to another acceptor thereby changing back from the reduced to the oxidised state. There has been much speculation in the literature on the routes the electrons take through the protein on their way to and from the Cu. For the Azu's the idea of two ports of entry/exit has circulated in the literature for a long time. One et relay would encompass the Cu ligand His-46 and residue His-35, which is in Van der Waals contact with His-46.[17-19] The other possible pathway would run through His-117.[19,20] This residue is the only copper ligand that protrudes through the protein's surface and makes contact with the surrounding solvent. Recent experiments with site-directed mutants have made it highly likely that the electrons leave and reach the Cu via His117, and that the "His-35 pathway" is not operative.[21,22] This holds for the electron self exchange (ese) reaction as well as for the et with physiological partners like cytochrome c_{551} and nitrite reductase. An interesting detail that appears from the X-ray structure, is that the protein surface at the location of His-117 exhibits a shallow depression.[1,6] This depression is filled by a water molecule that is connected by hydrogen bonds to His-117 and the protein backbone. Indirect NMR evidence indicates that also in solution this water molecule is immobilised to some extent with respect to the protein frame work.[8] On the basis of these observations it has been suggested that this water molecule has a functional role in the transfer of electrons.[6]

Also for Pc's the existence of two et routes has been postulated, one running through His-87, the analogue of His-117 in the Azu's, and one connecting the Cu with the "acidic patch" at the "Eastern side" of the protein.[23] The latter path would consist of the residues Tyr-83 and Cys-84, Cys-84 being a copper ligand. In this case there is experimental support for the existence of both routes. Electron transfer between Pc and cytochrome *f* appears to occur via the hydrophobic patch around His-87,[24] while the oxidised photosystem P700[+] appears to use the acidic patch to accept electrons from reduced Pc.[25]

Amc from *T. versutus* also contains a histidine (His-96) in a position similar to the positions taken by His-117 and His-83 in Azu and Pc, respectively. Also in Amc this histidine appears crucial for et.[13] An intriguing question which has not been addressed yet, is whether Amc, like Pc, has a second et route. The second route might encompass Phe-92 and Cys-93, the analogues of Tyr-83 and Cys-84 of Pc.

Lowering of the pH below 7 results in a decreased et rate of Pc and Amc. The origin of this reduction must be sought in the instability of the Cu-site in reduced Pc and Amc, which appears to be connected with the lability of the histidine ligand at the surface. Upon lowering of the pH this histidine becomes protonated, dissociates from the Cu and adopts a new conformation. The process has been nicely documented by Freeman and co-workers.[11] The kinetics of this conformational switch in Amc have been investigated by NMR. The dissociation of the histidine results in a change of the Cu coordination from four-fold to three-fold. The protein in the oxidised form, with the Cu in the 4-coordinated state, is insensitive to pH down to pH 4.[13] Reduction of the protein at low pH is accompanied by a change in the active site structure (4-coordinate Cu in the oxidised form, 3-coordinate in the reduced form). Presumably this increases the barrier for et and causes a drop in the et rate at low pH. In Azu, on the other hand, the Cu-site structure is insensitive to pH both in the reduced and in the oxidised protein. In the following section differences in the 3D structure of the proteins, especially of their Cu-sites, are delineated. Attention will be focused on the "axial ligands" in the Cu sites of the Azu's, and on the lability of one of the histidine ligands in Pc and Amc.

4. Cu-site structure

In the early reports on the structures of Azu and Pc the coordination of the Cu was described as distorted tetrahedral.[2,10] An N_2SS' donor set derives from the N^δ-atoms of two histidines, the S^γ-atom of a cysteine, and the S^δ-atom of a methionine (see figure 2). Later on it was noticed that in Azu's the backbone carbonyl oxygen atom from Gly-45 might belong to the coordination sphere of the Cu. The Cu site would then possibly be better described as trigonal bi-pyramidal with three strong in-plane ligands deriving from His-46, Cys-112 and His117, and two weak axial ligands deriving from Gly-45 and Met-121.[1] Indeed, it was found that the Cu almost resides in the plane formed by the N_2S (His$_2$Cys) donor set and that it is only slightly displaced out of this plane (~0.1 Å) towards the Met-121 sulfur. However, the long Cu-O bond (≥3 Å) precludes any covalent character in the Cu-Gly45 interaction, while the Cu-S(Met121) bondlength (2.6-3.1 Å) might allow for some covalency. In the Pc's, on the other hand, the Cu-site cavity below the N_2S plane, distal from the Met-ligand, is closed. In fact the distance between the Cu and the carbonyl that is homologous to the Gly-45 carbonyl in the Azu structure, amounts to

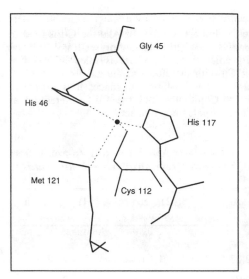

>3.9Å. Both in Pc and Amc, therefore, distorted tetrahedral seems the appropriate description of the Cu-site geometry. At this point it is worthwhile to see how the axial interactions distinguish the Cu site in the Azu's from those in the Pc's and Amc's.

Figure 2. Cu-site of azurin from *Pseudomonas aeruginosa.*[4,6] The ligands His-46, Cys-112, His-117 and Met-121 have been indicated by labels. Also the backbone carbonyl of Gly-45 is indicated. The black circle represents the Cu.

(i) Axial interactions

In the early attempts to express the P.a. Azu encoding gene in *Eschericia coli* the azurin appeared to be contaminated with a colorless species that was dubbed azurin* and later appeared to be Zn-azurin.[26] Resolution of the 3D-structure by means of X-ray diffraction techniques revealed a regular metal site in which the Zn, however, occupies a position which is clearly different from the position taken up by the Cu in the holo-azurin (see Table 1).[3] Instead of being located on the proximal side of the N_2S plane (the side at which Met-121 is placed) the Zn now occurs at the distal side of this plane and is displaced in the direction of the oxygen of carbonyl-45. The carbonyl in its turn has moved towards the Zn. The Met-121 S^δ clearly does not participate in the Zn coordination (see Table 1), and the metal coordination has changed from trigonal bi-pyramidal in Cu-azurin to distorted tetrahedral. This illustrates that carbonyl-45, although not binding covalently to the Cu, is not necessarily an innocent by-stander in terms of ligand-metal interactions. Coined in the language of "hard" and "soft" metals and ligands, the harder Zn (as compared to Cu) prefers the harder O of carbonyl-45 as a ligand over the softer S of Met-121. Apparently, since Zn^{2+} is iso-electronic with Cu^+, the difference of a single positive charge is sufficient to change the coordination geometry. The effect is all the more remarkable since changing the charge on the Cu from +1 to +2 has virtually no influence on the geometry of the Cu-site.

A second result revealing the influence of axial interactions on the metal site in Azu's, derived from the study of a mutant, M121Q azurin, in which the Cu-ligand Met-121 had been changed for a glutamine. M121Q Azu harbours a type 1 Cu site, albeit with spectroscopic properties that are more reminiscent of stellacyanin than of Azu.[7] Interestingly, the Cu-site in oxidised M121Q Azu again shows a distorted tetrahedral symmetry, this time with the Cu displaced towards the O^ε-atom of Gln-121 (see Table 1). Apparently, in this case the interaction of the Cu with the O^ε-121 prevails over the Cu-O45 interaction. Reduction of M121Q Azu is accompanied by a sizeable change in Cu-ligand bonding distances (Table 1), the reduced site being best described as approximately linearly 2-coordinated. In line with this the ese rate drops by two orders of magnitude as compared to wild type Azu. Apparently, the change in Cu-site geometry upon reduction considerably increases the kinetic barrier for et.

The results presented so far create the impression that in Azu the Cu is held in place in the N_2S-plane by three strong ligands (the N_2S-donor set), and that the position of the Cu in a direction perpendicular to the plane is ruled by a subtle balance between the interactions of the Cu with the atoms at the axial positions (O-45 and S^δ-121 or O^ϵ-121). On the basis of the Cu-ligand distances it is logical to consider the Cu-O interaction as wholly Coulombic and the Cu-S interaction as partly covalent, partly Coulombic (induced dipole) in character.

Table 1. Metal ligand bond distances in Å (± 0.05 Å) for Cu- and Zn-azurin from *Pseudomonas aeruginosa*[4,6] and M121Q Cu-azurin from *Alcaligenes denitrificans*.[7]

	Cu-azurin	Zn-azurin	M121Q azurin oxidised	M121Q azurin reduced
Gly45 O	2.97	2.32	3.37	3.34
His46 N^δ	2.11	2.07	1.94	1.98
His117 N^δ	2.03	2.01	2.05	2.68
Cys112 S^γ	2.25	2.30	2.12	2.09
Met121 S^δ	3.15	3.38	--	--
Gln121 O^ϵ	--	--	2.29	2.75
Cu-N_2S^*	0.10	-0.15	-0.27	-0.20

*   *Distance from Cu to N_2S plane. Positive/negative values correspond with location of Cu at the proximal/distal side of the N_2S plane*

Support for this view comes from a series of molecular dynamics (MD) simulations aimed at testing various force fields for the Cu in oxidised and reduced Azu.[27] The simulations were inspired by the lack of a proper set of parameters for Cu in the customary MD force fields. A metal force field was sought that would allow for a stable Cu-site in Azu during an MD run. An important criterion in judging the validity of a parameter set was whether the Cu would stay on the proper side of the N_2S plane during an MD run. Lennard-Jones parameters were chosen such that the Cu site remained stable during an energy minimisation (EM) search and a restrained MD run. A subsequent <u>unrestricted</u> MD run showed that a second lower minimum for the Cu occurred at the distal side of the N_2S plane. Further adaptations appeared necessary, therefore, including the introduction of a number of weak Cu-X-Y bending modes and improper dihedrals in the N_2S plane. However, the Coulomb forces between O-45 and the Cu kept pulling the Cu to the distal side of the N_2S plane. Eventually, to obtain a stable Cu site it appeared essential to diminish the strength of the Coulomb interactions inside the Cu cavity. To this end the net charge on the metal (Cu + ligands) was reduced to +0.58 and 0.0 in the Cu^{2+} and Cu^+ state, respectively. Moreover, a large part of the nominal charge on the metal had to be distributed over the ligands. For the Cu^{2+} state good results were obtained with charges of +0.40 on the Cu, +0.23 on the histidines and -0.28 on the cysteine. For the reduced protein these numbers amounted to +0.18, +0.14 and -0.46, respectively. In addition the Met-121 side chain had to be slightly polarised and the magnitude of the carbonyl-45 dipole had to be reduced by lowering the charges on the oxygen and the carbonyl carbon from ± 0.38 to ± 0.20.

According to this approach the change in charge accompanying a change in oxidation state is buffered by the Cu ligands. The coordination shell of the Cu acts as a charge sink which might explain why the metal coordination has been found to be insensitive to the oxidation state of the Cu.[28,4] The results also demonstrate that the interaction of the Cu with carbonyl-45, although mainly Coulombic in character, can be substantial. In this respect the Cu-sites in the Azu's differ significantly from those of plastocyanins and amicyanins, although the functional relevance of this difference remains difficult to grasp. Possibly the carbonyl dipole is functional in promoting the incorporation of the metal during the *in vivo* formation of the holo-protein.

(ii) His-117

As pointed out above the mobility of His-87 and His-96 in Pc and Amc, respectively, is missing in the case of His-117 in Azu. We have studied the importance of His-117 for the Cu-site by replacing this residue with a glycine. It appeared possible to fill the hole in the protein mantle created in this way, by adding various ligands. The Cu site of the oxidised Azu could be succesfully mimicked with a substituted imidazole (Imz) as judged from spectroscopic features.[29,30] The association constant for the reaction between H117G and Imz, however, amounts to 10^3 M^{-1}, which is fairly low. The association can be improved by making use of the hydrophobic properties of the surface area around His-117. For instance, when an alkane spacer is provided with Imz groups at both ends, dimers with H117G Azu are formed with a much higher association constant.[31] The association is promoted because the azurins in the dimer presumably align their hydrophobic patches while binding to the spacer.

When the H117G+Imz azurin is reduced the blue copper site is irreversibly lost. The site can neither be reconstituted by adding Cu nor reoxidised. A likely explanation is that Imz drops out of its pocket upon reduction of the Cu, analogous to the behaviour of His-83 and -96 in Pc and Amc. Contrary to the latter two cases, however, the resulting site is instable possibly because it has acquired type 2 character.[29,30] This may promote chemical activity of the electron on the reduced Cu. The details of this reaction are being studied at the moment.

The findings outlined here illustrate the importance of the covalent link between the Imz side chain and the protein frame to maintain the integrity of the Cu site in Azu. The side chain is fixed and is unable to rotate the way it does in Pc and Amc. Apparently, it is not the strength of the binding to the Cu that is responsible for this feature of Azu, since the side chain is easily lost when the link is cut. Rather, we think, it is the protein matrix that inhibits rotation of His-117 in Azu and that does allow for such a rotation in Pc and Ami. This idea is supported by structural studies on the apo-forms of Pc and Azu.[5,32] In apo-Pc the protein matrix provides room enough for His-83 to assume a position in which the side chain has rotated by about 180° around the C^β-C^γ-bond. In apo-Azu, on the other hand, the His-117 ring can move outwards in the direction of the protein surface, but remains co-planar with the position it has in the holo-Azu. The cavity in which the His side chain is packed in Azu only allows a sliding motion like that of a credit card in a credit card slot.

5. Conclusion

The first conclusion from the results presented above is that His-117 in Azu protects the Cu site from unwanted side reactions. When the side chain is removed the copper site becomes unstable in the reduced form. Experiments with the enzyme ethyl-phenylmethylene-hydroxylase (EPMH) further illustrate the protective role of His-117.[33] EPMH is able to oxidise p-ethylphenol (EP). The electrons produced in this reaction are taken up by the EPMH and can subsequently be accepted by Azu. In the presence of EP, EPMH reacts with about equal efficiency with H117G Azu and H117G+Imz. However, when the enzyme is absent it appears that the EP is able to attack the H117G directly, albeit slowly, unless the Cu site is protected by an Imz group. Evidently, to maintain the integrity and activity of the metal site in Azu it is important that His-117 is not as mobile as His-87 and His-96 in Pc and Amc, respectively. This is achieved by packing the His-117 side chain between bulky side chains (Met-13 and Phe-114). The mobility of His-117 is limited to a sliding movement which possibly assists in the uptake of Cu.[5] Why the metal site in the H117G mutant is vulnerable after reduction and whether the accessibility of O-45 in the Cu site cavity plays a role in this is subject of further experimentation.

The second conclusion is that His-117 and its analogues in Pc and Amc have a functional role in the conduction of electrons to and from the Cu. The dissociation of this histidine from the reduced Cu at low pH in Pc and Amc, with a concomitant decreased efficiency of et, may have significance for the functioning of the corresponding et chains *in vivo*. Enhanced energy metabolism lowers the pH in the periplasmic or thylakoid space. Coupling of the et efficiency of one of the links in the et chain to the pH may constitute a feed-back mechanism to control the rate of energy production in the cell.

Finally, the results reported here allow for some speculation about the evolutionary relationship between the blue copper proteins dealt with above. It is generally accepted that the cyanobacteria, and the chloroplasts in green plants and algae, which nearly all harbour plastocyanins in their photosynthetic et chains, derive from a common procaryotic ancestor equipped with some kind of photosynthetic machinery. Amicyanins, on the other hand, have been found until now only in fairly primitive methylotrophs. These organisms are non-phototrophic. Yet, the secondary structure of the amicyanins as well as the structural and mechanistic details of their Cu-sites clearly point to an evolutionary relationship with the Pc's. Indeed, the methylotrophs in which amicyanins have been found so far, are often considered to be related to photosynthetic bacteria. Plastocyanins and amicyanins, therefore, may derive from a common blue copper protein that once occurred in a phototrophic ancestor.

Assuming that the Azu's, Pc's and Amc's derive from a common ancestral potein, the structural and mechanistic traits dealt with above seem to justify the conclusion that the branching point between the Pc's and Amc's on the one hand and the Azu's on the other hand lies before the point where Pc and Amc began to diverge. This leads to the question what the characteristics of the prototypical type 1 site might have been. The study of the recently described type 1 Cu sites in nitrite reductases[34] may shed some new light on this subject.

References

1.   E.N. Baker, J. Mol. Biol., 1988, 203, 1071
2.   E.T. Adman and L.H. Jensen, Isr. J. Chem., 1981, 21, 8
3.   H. Nar, R. Huber, A. Messerschmidt, A.G. Filippou, M. Barth, M. Jaquinod, M. van de Kamp and G.W. Canters, Eur. J. Biochem., 1992, 205, 1123
4.   H. Nar, A. Messerschmidt, R. Huber, M. van de Kamp and G.W. Canters, J. Mol. Biol., 1991, 221, 765
5.   H. Nar, A. Messerschmidt, R. Huber, M. van de Kamp and G.W. Canters, FEBS Lett., 1992, 306, 119
6.   H. Nar, A. Messerschmidt, R. Huber, M. van de Kamp and G.W. Canters, J. Mol. Biol., 1991, 218, 427
7.   A. Romero, C.W.G. Hoitink, H. Nar, A. Messerschmidt, R. Huber and G.W. Canters, J. Mol. Biol., in press
8.   M. van de Kamp, G.W. Canters, S.S. Wijmenga, A. Lommen, C.W. Hilbers, H. Nar, A. Messerschmidt and R. Huber, Biochemistry, 1992, 31, 10194
9.   C.W.G. Hoitink, unpublished results
10.   J.M. Guss and H.C. Freeman, J. Mol. Biol., 1983, 169, 521
11.   J.M. Guss, P.R. Harrowell, M. Murata, V.A. Norris and H.C. Freeman, J. Mol.

Biol., 1986, <u>192</u>, 361

12. J.M. Moore, C.A. Lepre, G.P. Gippert, W.J. Chazin, D.A. Case and P.E. Wright, <u>J. Mol. Biol.</u>, 1991, <u>221</u>, 533

13. A. Lommen and G.W. Canters, <u>J. Biol. Chem.</u>, 1990, <u>265</u>, 2768

14. A. Lommen, S. Wijmenga, C.W. Hilbers and G.W. Canters, <u>Eur. J. Biochem.</u>, 1991, <u>201</u>, 695

14a. J. Van Beeumen, S. Van Bun, G.W. Canters, A. Lommen and C. Chothia, <u>J. Biol. Chem.</u>, 1991, <u>266</u>, 4869

14b. L. Chen, R. Durley, B.J. Poliks, K. Hamada, Z. Chen, F.S. Mathews, V.L. Davidson, Y. Satow, E. Huizinga, F.M.D. Vellieux and W.G.J. Hol, <u>Biochemistry</u>, 1992, 4959

15. A. Kalverda, A. Lommen, S.S. Wymenga, C.W. Hilbers and G.W. Canters, <u>J. Inorg. Biochem.</u>, 1991, <u>43</u>, 171

16. I. Chen, R. Durley, B.J. Poliks, K. Hamada, Z. Chen, F.S. Mathews, V.L. Davidson, Y. Satow, F.M.D. Vellieux and W.G.J. Hol, <u>Biochemistry</u>, 1992, <u>31</u>, 4959

17. O. Farver, Y. Blatt and I. Pecht, <u>Biochemistry</u>, 1982, <u>21</u>, 3556

18. O. Farver and I. Pecht, <u>Isr. J. Chem.</u>, 1981, <u>21</u>, 13

19. A.G. Lappin, M.G. Segal, D.C. Weatherburn, R.A. Henderson and A.G. Sykes, <u>J.Amer. Chem. Soc.</u>, 1979, <u>101</u>, 2302

20. C.M. Groeneveld and G.W. Canters, <u>Eur. J. Biochem</u>, 1985, <u>153</u>, 559

21. M. van de Kamp, M.C. Silvestrini, M. Brunori, J. van Beeumen, F.C. Hali and G.W. Canters, <u>Eur. J. Biochem.</u>, 1990, <u>194</u>, 109

22. M. van de Kamp, R. Floris, F.C. Hali and G.W. Canters, <u>J. Amer. Chem. Soc.</u>, 1990, <u>112</u>, 907

23. A.G. Sykes, <u>Adv. Inorg. Chem.</u>, 1991, <u>36</u>, 377

24. S. He, S. Modi, D.S. Bendall and J.C. Gray, <u>EMBO J.</u>, 1991, <u>10</u>, 4011

25. M. Nordling, K. Sigfridsson, S. Young, L.G. Lundberg and O. Hansson, <u>FEBS Let.</u>, 1991, <u>291</u>, 327

26. M. van de Kamp, F.C. Hali, N. Rosato, A. Finazzi Agro and G.W. Canters, <u>Biochim. Biophys. Acta</u>, 1990, <u>1019</u>, 283

27. G.W. Canters, W.F. van Gunsteren and H.C. Berendsen, unpublished results

28. W.E.B. Shepard, B.F. Anderson, D.A. Lewandowski, G.E. Norris and E.N. Baker, <u>J. Amer. Chem. Soc.</u>, 1990, <u>112</u>, 7817

29. T. den Blaauwen, M. van de Kamp and G.W. Canters, <u>J. Am. Chem. Soc.</u>, 1991, <u>113</u>, 5050

30. T. den Blaauwen and G.W. Canters, <u>J. Am. Chem. Soc.</u>, 1993, in press

31. T. den Blaauwen and G. van Pouderoyen, unpublished results

32. T.P.J. Garrett, D.J. Clingeleffer, J.M. Guss and H.C. Freeman, <u>J. Biol. Chem.</u>, 1984, <u>259</u>, 2822

33. A.C.F. Gorren, unpublished results

34. E.T. Adman, <u>Curr. Opin. Struct. Biol.</u>, 1991, <u>1</u>, 895

Bacterial Plasmid Resistances to Copper, Cadmium, and Zinc

Simon Silver[1,2], Barry T. O. Lee[2,3], Nigel L. Brown[2,3], and Donald A. Cooksey[4]

[1] UNIVERSITY OF ILLINOIS, CHICAGO, IL 60680, USA
[2] UNIVERSITY OF MELBOURNE, PARKVILLE, VICTORIA 3052, AUSTRALIA
[3] UNIVERSITY OF BIRMINGHAM, BIRMINGHAM B15 2TT, UK
[4] UNIVERSITY OF CALIFORNIA, RIVERSIDE, CA 92521, USA

1   INTRODUCTION

This Conference on "Copper and Zinc Triads" includes Ag, Au, Cd and Hg, as well as Cu and Zn. For cations of five of these six elements (all except gold), there are highly specific bacterial resistance systems and the task of our laboratories has been to explain the basic biochemical mechanisms of these resistances as well as the molecular biology and genetics that determine and govern them. Much recent progress has been achieved on copper, zinc, and cadmium resistance systems.[1-3] We will limit this report to new findings on bacterial copper resistance (which leads to a picture substantially different from one only a few months ago[1,2,4,5]) and to new progress on cadmium and zinc resistance systems. Some of this work was recently summarized.[1,3] Although not discussed here, mercury resistance is the most thoroughly studied of all bacterial heavy metal resistance systems.[1,4,6-8] An additional review of mercury resistance is not needed. Bacterial resistance to silver occurs[6] but very little new work has been done in recent years and a basic understanding of the mechanism of silver resistance is lacking.

Copper and zinc resistance mechanisms present the cell with a special problem because these are essential nutrients at low levels, as well being toxic at high levels. Therefore, the resistance mechanisms and their genetic controls are responsive to the requirement to accumulate cations at trace levels and at the same time to reduce cytoplasmic concentrations from potentially toxic levels.[2,4] Careful regulation appears to result from interplay between chromosomally-determined cation uptake systems and plasmid-determined efflux systems, with regulatory genes on both the chromosome and the plasmid responding in a gradual fashion to increasing cation levels.[2,4]

2   PLASMID-DETERMINED COPPER RESISTANCE SYSTEMS

Copper resistance has been observed frequently in bacteria. What were thought to be different plasmid copper resistance determinants in <u>Pseudomonas</u> <u>syringae</u> and in

<u>Escherichia coli</u> have recently been shown by DNA sequencing analysis to be related. However, the basic mechanisms of the <u>cop</u> system on a <u>P. syringae</u> plasmid and of the <u>pco</u> system in <u>E. coli</u> are not as well understood as we would wish. Overall, both systems include (a) genes on the bacterial chromosome that govern copper uptake (and probably intracellular sequestration and efflux of copper) and (b) genes from copper resistance plasmids (which function similarly to their chromosomal counterparts). There is an interdependency between products of chromosomal genes and plasmid genes. This assures the meeting of normal cellular needs for low levels of copper, and at the same time affords intracellular protection from redox damage that free copper ions might cause and also from overly high cellular copper levels when normally toxic levels are added to resistant cells. As well as having gene products that interact for the functions of the copper resistance systems, there are regulatory gene(s) on the plasmid whose products interact with products of chromosomal regulatory genes, both <u>E. coli</u> and <u>Pseudomonas</u>.

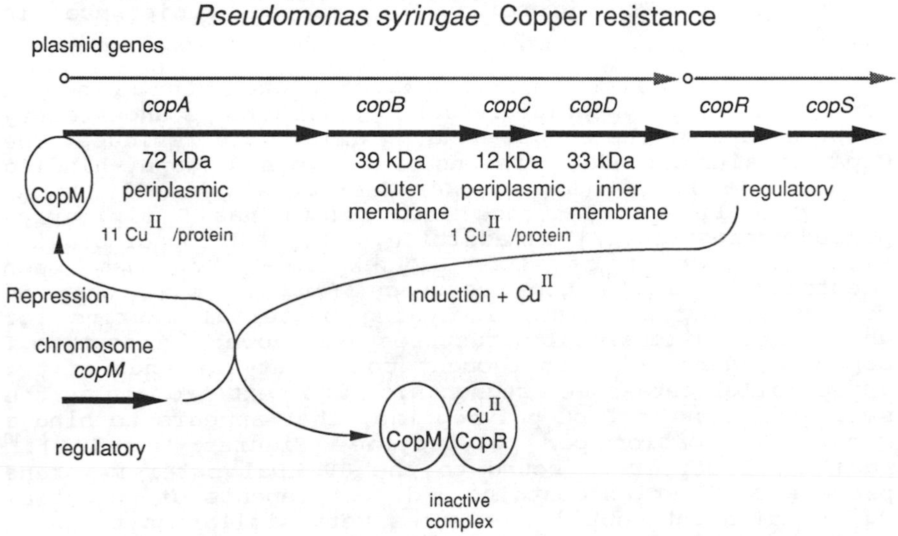

Figure 1 The genetic control of the <u>Pseudomonas</u> copper resistance system (updated from ref. 1 with permission).

Figure 1 shows the current picture of the genetic determinants of copper resistance on a plasmid of <u>P. syringae</u> isolated from plants sprayed with copper salts as a fungicide.[5,9-11] Additional and similar copper resistance determinants are quite widespread in nature, wherever copper-salts were being sprayed for plant pest control.[12]

Four structural genes, <u>copA</u>, <u>copB</u>, <u>copC</u> and <u>copD</u>,are found on the plasmid.[9] Their products determine two periplasmic proteins (CopA and CopC) plus an inner (CopD) and an outer (CopB) membrane protein (Figure 2).[10] All four of the predicted <u>Pseudomonas</u> Cop polypeptides start with canonical membrane signal sequences.[9] Cleavage of the CopA, CopB, and CopC signal peptides at the predicted positions was directly measured.[10] Mutations in any of these genes leads to copper sensitivity, and mutant strains that are CopA⁻ CopB⁻ CopC⁺ CopD⁺ are even more copper sensitive than strains without plasmids.

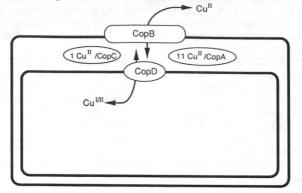

Figure   2   The   proteins   for   copper   resistance   in *Pseudomonas*.

CopA is a 72 kDal periplasmic protein that binds 11 Cu^{2+} atoms per polypeptide (Figure 2).[10] There is uncertainty about nature of the copper-binding amino acid residues. The CopA contains 5 tandem versions of a proposed copper-binding octapeptide motif (Asp-His-Ser-Xaa-Met-Gln/Ala-Gly-Met).[9] [The *E. coli* PcoA sequence (see below) has 5 similarly-placed octapeptides.] In addition, a "type I copper binding site" consisting of $His-X_{48}-Cys-X_4-His-X_4-Met$ has been identified[13] in the CopA sequence (and in more than 50 copper-binding proteins, including bacterial azurins for which three-dimensional structures are known). This type I copper-binding motif is thought to lie at the ends of two antiparallel beta-sheet segments.[13] The next protein is the small periplasmic CopC polypeptide, that appears to bind a single Cu^{2+} cation per polypeptide (Figures 1 and 2).[10] Additional copper is bound to the 39 kDal outer membrane protein CopB, which contains 5 direct repeats of potential Cu^{2+}-binding octapeptide sequences very similar to those in CopA.[5,9] Crude outer membrane fractions from CopB-containing cells were blue with bound copper, but the copper was removed during subsequent purification of CopB protein.[10] We suggest that the outer membrane CopB protein may be involved in cross-membrane transport as well as copper binding. In summary, there are two periplasmic copper-binding proteins, CopC and CopA, plus the copper-binding outer membrane protein CopB (Figure 2), but basically, we do not know how Cu^{2+} is bound by these proteins. There is another known outer membrane copper-translocating protein, NosA, involved in inserting Cu^{2+} into the periplasmic copper-containing nitrous oxide (N_2O) reductase of *Pseudomonas stutzeri*.[14,15] The NosA protein also binds copper. However, the form of the copper in NosA may be quite different from Cu^{2+} associated with CopB. The NosA protein is much larger than CopB and shows no sequence homology to CopB.[14]   There also is no significant sequence homology between the periplasmic components of the copper resistance and the nitrous oxide reductase copper-protein.

The fourth <u>Pseudomonas</u> copper resistance polypeptide, CopD, is a 33 kDal inner membrane protein (Figures 1 and 2). Other than a role in inward and/or outward copper transport suggested by its cellular location, the predicted amino acid sequence of CopD gives no hint as to its protein function.

In addition to the plasmid <u>cop</u> gene products conferring copper resistance, related Pseudomonads can develop chromosomal mutations to copper resistance, which may result from "activation" of chromosomal <u>cop</u> homologues that were previously found by DNA/DNA hybridization.[11,12]

The copper-resistant cells turn bright blue during growth in copper-containing medium. The bluish copper-colour of the medium decreases. Copper accumulation of induced resistant cells was approximately 100-times higher than that of unexposed cells.[10] In fully induced cells, CopA constituted 3% of the total cellular protein and accounted for the surface copper accumulation and blue colour of <u>cop</u>-containing cells grown in high Cu^{2+}. From these results, Cooksey[5,10,11] proposed that the mechanism of Cu^{2+} resistance is periplasmic binding and extracellular sequestration of copper cations, decreasing free extracellular Cu^{2+} and protecting the cytoplasm from exposure to toxic levels. This resistance mechanism involving bio-accumulation is unlike other divalent cation resistance mechanisms (see below) which generally involve lower (not higher) cellular accumulation.

The <u>Pseudomonas</u> copper resistance determinant is regulated at a promoter site just upstream from the first gene, <u>copA</u>,[1,15] and all four genes are transcribed as a single unit (Figure 1). Using a reporter gene, Mellano and Cooksey[16] demonstrated that as little as 1 μM Cu^{2+} functioned as inducer and that other divalent cations (Ca^{2+}, Mn^{2+}, Fe^{2+}, Zn^{2+}, Cd^{2+}, Hg^{2+} and Pb^{2+}) did not induce the <u>cop</u> system. With a <u>pco</u>-<u>lux</u> fusion plasmid for the <u>E</u>. <u>coli</u> <u>pco</u> system (see below), D. Holmes and S. Gangulli (personal communication) found high specificity for copper (with Ni^{2+}, Co^{2+}, Zn^{2+} and Cd^{2+} showing less than 1% of the Cu^{2+} response). The start site of the <u>cop</u> transcript has now been defined but canonical <u>E</u>. <u>coli</u>-like or <u>Pseudomonas</u>-like transcriptional signals are missing. Only the first 200 base pairs of the <u>cop</u> operon are needed for inducible expression of gene fusions[16] and the hypothesized trans-acting "CopM" repressor function appears to be provided by a <u>cop</u> operon homolog found on the chromosomes of some but not all plant pathogenic <u>Pseudomonas</u> strains, sensitive as well as resistant isolates. Cooksey[5] has identified three potential components in regulation of the <u>cop</u> copper-resistance determinant (Figure 1). There is the chromosomal determinant of a DNA binding protein, which is found only in related <u>Pseudomonas</u> strains and that we will tentatively call CopM. This protein activity was recognized by gel mobility shift with a DNA fragment containing the specific cop promoter region. When the protein extract was prepared from cells grown on high Cu^{2+}, the protein-DNA association was disrupted by a plasmid-encoded function, CopR plus CopS[5] (Figure 1), encoded by two separately transcribed genes that follow copD (Figure 1). DNA sequence analysis[5] indicates the presence of these two genes, which are members of the same two-component family as pcoR and pcoS of the E. coli pco system (see below).

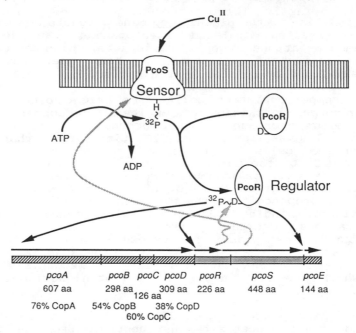

pcoA	pcoB	pcoC	pcoD	pcoR	pcoS	pcoE
607 aa	298 aa	126 aa	309 aa	226 aa	448 aa	144 aa
76% CopA	54% CopB	60% CopC	38% CopD			

Figure 3   The regulation of the E. coli copper resistance determinant (revised and updated from ref. 2).

A second system for plasmid-encoded copper resistance appeared in E. coli isolated from a pig fed copper salts as a growth stimulant.[2,17-19] Related plasmid resistance systems for copper have been identified in a variety of enteric bacteria isolated under similar circumstances.[20] This system was cloned and shown to govern an inducible resistance that also involves copper binding and transport components.[2, 17-19] Whereas previously, a single regulatory gene pcoR and a minimum of three structural genes had been identified by mutation and complementation analysis, recent DNA sequencing of the plasmid pco system defines a region of seven structural genes, pcoA-E, pcoR and pcoS (Figure 3). The pco region contains four structural genes pcoABCD (previously part of a single long pcoA complementation unit) that show amino acid sequence homology to the four Pseudomonas plasmid genes copABCD. The evidence for the relationship between pco and cop currently rests solely with the DNA sequences: the amino acid sequence identities between the PcoA and CopA proteins is 76%, whereas PcoB and CopB are 54% identical at the amino acid level, PcoC and CopC 60%, and PcoD and CopD

38%, respectively (Figure 3). PcoA and PcoD have His- and
Met-containing octapeptide motifs related to those described
above for Cop polypeptides.

Following <u>pcoD</u> comes the regulatory gene <u>pcoR</u>
previously identified at that position.[2,17-19] <u>pcoR</u> was
initially shown to determine a trans-acting regulatory
factor[18]; its predicted amino acid sequence shows strong
homologies to members of the second "transducer" components
of the large family of two component, autophosphorylating
(sensor protein)-transphosphorylating (transducer/ DNA
binding protein) regulatory systems.[21] PcoR is homologous to
CopR (61% amino acid identities) and then to the copper-
sensing regulatory protein CutR from <u>Streptomyces lividans</u>[22]
(41% identities). PcoR is also related (about 30% amino acid
identities) to other DNA-binding transducer proteins
including PhoB and KdpE of the Pst phosphate and Kdp
potassium transport systems (Figure 4 and refs. 1, 21).

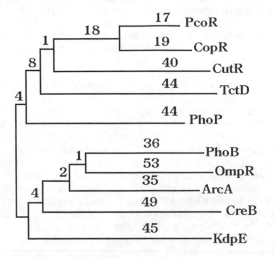

<u>Figure 4</u>   The relationships between the sequences of the
copper regulatory proteins (CopR from <u>P. syringae</u>, PcoR from
<u>E. coli</u> and CutR from <u>S. lividans</u>) and other DNA binding
proteins of the transmitter class . The numbers are in
arbitrary distance units based on similarities of sequences,
both identical and related amino acids.

After <u>pcoR</u>, the next open reading frame in the DNA
sequence has been named <u>pcoS</u>, by homology to the sensor
components of the two component systems.[1,21] The predicted
sequence of PcoS indicates it to be a membrane protein, as
is frequently the case with such auto-phosphorylating
proteins. From sequence homologies, it seems very likely
that the kinase activity of PcoS will auto-phosphorylate on
a conserved His_{257} residue and that the phosphate will be
subsequently transferred to Asp_{52} on PcoR, as indicated in
Figure 3. Phosphorylated PcoR is then predicted to activate

several operator/promoter regions in the pco determinant,
increasing transcription of mRNA. Four transcripts are
indicated on Figure 3, but the actual number and their
starting points are not yet known. The existence of four
complementation groups and a range of pco-lacZ fusions[18] lead
to the suggestion that there are several transcripts
involved, rather than the single transcript indicated for
the cop system in Figure 1. Whether pco and cop will differ
in this way is still uncertain. In the earlier
complementation analysis[18], the region currently assigned to
the single open reading frame pcoS contained two
complementation groups that were then called pcoB and pcoC.
It is possible that the C-end of pcoS may determine a
soluble 26 kDal polypeptide that binds Cu^{2+}.[2,4,17,18] A
candidate for an internal (within pcoS) in-frame start site
has been identified in the DNA sequence. If these
preliminary findings hold up, it may be that internal Cu^{2+}
as well as extracellular Cu^{2+} are sensed by PcoS. Finally,
it is uncertain whether regulation will be positive (as is
frequently the case for two-component phospho-protein gene
regulation) or negative as currently thought to be the case
for cop[5] (Figure 1). However, the recent DNA sequencing
results make it more likely that cop and pco will share
basic genetic regulatory mechanisms as well as biochemical
mechanisms.

Previous models of the Pco copper resistance system
suggested the existence of proteins involved in
intracellular Cu^{2+} binding and Cu^{2+} efflux.[1,2,4,19] Whether
proteins with these roles will be found is still an open
question for this currently preliminary picture of bacterial
copper resistance. However, chromosomal genes with such
roles have also been tentatively identified.[18,23]
Understanding of these chromosomal genes, in which mutations
lead to copper-hypersensitivity, is currently less even than
that for plasmid-determined copper resistance systems. In E.
coli, six mutations leading to increased Cu^{2+} sensitivity
were obtained.[16,22] Each mutation mapped in a separate
chromosomal location from the others. Only region one has
been cloned and sequenced.[24] The cutE gene has been located
at minute 15.5 (kilobase position 706) on the standardised
E. coli physical map and its DNA sequence has been
determined.[24] [Note that Cut has been used for Cu^{2+}
transport-related genes for for the E. coli chromosomal
system[2,4] and for probably different functions in
Streptomyces lividans copper metabolism.[22]] As a soluble
protein, CutE has been assigned an intracellular Cu^{2+}-binding
role. Prior to gene isolation or sequencing, CutF has been
assigned a similar position, whereas CutA and CutB are
thought to be involved in Cu^{2+} uptake (as membrane transport
proteins) and CutC and CutD are thought to be involved in
the efflux of excess Cu^{2+}, perhaps together with pco gene
products.[1,2,4,17] P. syringae also has cut-type chromosomal
mutations that lead to copper sensitivity. There is a recent
report with still another bacterium, Enterococcus hirae,
that a P-class ATPase[1,25], previously thought to be involved
in transport of potassium, is probably involved in Cu^{2+}
efflux,[26] as pictured for some cut genes of E. coli. E. hirae

has contiguous genes for two P-class ATPases and disruption of these genes leads to Cu^{2+} sensitivity.[26] Of the six cut mutations conferring Cu^{2+} sensitivity, preliminary experiments show some with changes in the uptake or retention of $^{64}Cu^{2+}$.

Finally, the chromosomal copper transport system is hypothesized also to be regulated by a two component system, cutR and cutS.[17,18,23] The possibility that there are homologous two-component regulatory systems on both the plasmid pcoRS and on the chromosome cutRS opens up the potential of interaction (cross talk)[21] between functionally related cut and pco regulatory genes. Repeatedly, copper transport and sensitivity in bacteria appears to result from interactions between products of plasmid-encoded and chromosomally-encoded genes.

There is a newly-considered aspect of basic cellular metabolism that is expected to affect copper sensitivity and resistance, but that has been left out from earlier considerations. Small intracellular thiol-containing polypeptides may chelate copper and protect against cellular damage and sometimes perhaps deliver Cu^{2+} or Cu^+ from one cellular component to another. Preliminary growth experiments with E. coli mutants defective in glutathione or thioredoxin synthesis[27-29] indicate that these mutants are sensitive to high Cu^{2+}. Mutations affecting these cellular thiols might be expected to be found among the cut genes, whose phenotype is precisely that: sensitivity to added Cu^{2+}. Glutathione (the tripeptide, γ-glutamyl-cysteinyl-glycine; GSH) is the major intracellular thiol compound for bacterial cells, occurring in the millimolar range.[27] In addition, there are two related small di-thiol proteins, whose roles include coupling redox energy to cellular processes and maintaining the reduced state of other cellular thiol-containing proteins. These are thioredoxin[28,29] (a 12 kDal protein involved in ribonucleotide reduction and reduction of sulphur brought into the cell as sulphate) and glutaredoxin (a 9.5 kDal polypeptide, with related di-thiol sequence and functions). Mutants defective in thioredoxin have been found to be Cu^{2+} sensitive (preliminary data not shown). It is premature to make explicit models for the roles of these intracellular thiol peptides in Cu^{2+} uptake and sequestration. However, it is long overdue to consider these compounds in models of Cu^{2+} sensitivity and resistance.

3   CadA CADMIUM RESISTANCE

Cadmium resistance has been best characterized for the cadA resistance operon of S. aureus plasmid pI258.[1,32] Cadmium resistance results from lowered cellular Cd^{2+} accumulation which in turn comes from a plasmid-determined Cd^{2+} (and perhaps Zn^{2+}) energy-dependent efflux pump.[30] Recently, the net efflux system from intact Bacillus cells has been characterized in cell-free inside-out membrane vesicles, where Cd^{2+} uptake is plasmid-governed and ATP-dependent.[31] This system was initially identified as an ATPase from interpretation of the DNA sequence.[25,32] Recently

closely homologous systems have been found in two additional
bacteria: with an alkalophilic (high pH-growing) <u>Bacillus</u>[33]
and in the chromosomal determinant from an antibiotic-
resistant <u>S</u>. <u>aureus</u> (D.T. Dubin, personal communication).
Thus the CadA system promises to be wide- spread among Gram-
positive bacteria.

The cadmium resistance operon consists of two genes[32,34]:
the first, <u>cadC</u>, encodes a soluble protein of 122 amino
acids in length, and the second, <u>cadA</u>, encodes a 727 amino
acid membrane protein. The function of the highly charged
CadC protein is not known, but it is needed for full Cd^{2+}
resistance.[34] Highly homologous <u>cadC</u> genes have also been
found with the <u>Bacillus</u>[33] and chromosomal staphylococcal
systems (D.T. Dubin personal communication). CadC is also
needed for Cd^{2+} uptake by inside-out membrane vesicles.[31]

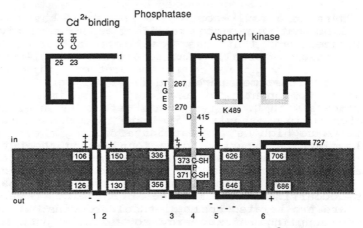

<u>Figure 5</u>   Functional model of the CadA Cd^{2+} efflux ATPase
of Gram-positive bacteria (from ref. 1 with permission).

The predicted CadA polypeptide sequence is strongly
homologous to the P-class ATPases, which are found in
membranes of all cells, of bacteria, animals and plants.[25]
The 727 amino acid CadA polypeptide sequence contains all of
the characteristic domains of the P-class ATPases and the
model in Figure 5 was constructed by analogy to the well-
studied Na^+/K^+- and Ca^{2+}- ATPases of animal cell membranes
and the muscle sarcoplasmic reticulum.   The sequence
contains a series of recognizable motifs and key conserved
residues:
        (a) The sequence starts with a hypothesized Cd^{2+}-
recognition region, including the Cys_{23} and Cys_{26} residues,
which are predicted to make initial contact with the Cd^{2+}
ion.   This region of the CadA amino acid sequence bears a
close sequence homology to the predicted Hg^{2+}-binding motifs
on two proteins of the mercury resistance system.[1,6,7]   The
chromosomal <u>Staphylococcus</u> system has an N-terminal repeat
of 70 amino acids for this Cd^{2+} binding motif (D.T. Dubin,
personal communication), just as the <u>Bacillus</u> mercuric
reductase has an N-terminal repeat of 79 amino acids
postulated to govern initial binding of Hg^{2+}. It appears as

if a "soft" metal (Cd^{2+} or Hg^{2+}) di-thiol binding motif has been recognized by comparing amino acid sequences from different DNA sequences.

(b) The next motif in the sequence is the first of 3 transmembrane hairpin structures, with closely-spaced, presumedly alpha-helical hydrophobic amino acid segments. Charged amino acids are located mostly-positive at the inner surface of the membrane hairpins and mostly-negative on the outer surface. The turn at the outer surface of the membrane between the first two membrane spans is very short (predicted to be only 4 amino acids). Overall approximately 20% of the CadA aminio acids are postulated to be in the membrane segment and almost 80% cytoplasmic.

(c) The third segment of the polypeptide chain consists of approximately 190 amino acids (Figure 5) that by homology to the other ATPases in this class has been assigned two roles. It has been considered to be a transduction "funnel" that may be involved in moving the trapped Cd^{2+} cation from its initial binding site to the inner membrane surface. Alternatively, this region may have phosphatase activity, removing the phosphate from its covalently bound position on Asp_{415}.

(d) The second membrane hairpin structure (from position 336 to 384 in the model in Figure 5) is thought to be the site of the cation transport channel across the membrane in all P-class ATPases.[25] There is a proline residue found in the comparable position in all P-class ATPases. In CadA, Pro_{372} is between Cys_{371} and Cys_{373}, which together with Cys_{23} and Cys_{26} are the only 4 cysteine residues in this cysteine-poor polypeptide.

(e) Following the channel, there is a large intracellular domain that includes an ATP-binding site (near Lys_{489}) and heptapeptide kinase segment (Asp_{415}-Lys-Thr-Gly-Thr-Leu-Thr$_{421}$) that is conserved in all P-class ATPases. Asp_{415} has been labelled with radioactive ^{32}Pi from γ-phosphate-labelled ATP by A. Linet and colleagues (personal communication).

(f) Following the third membrane hairpin structure that is highly conserved in P-class ATPases, the CadA sequence ends at Lys_{727}.

The CadC polypeptide sequence from plasmid pI258 is most closely related to that of CadC from the <u>Bacillus</u> strain[33] (83% identical amino acids) and less closely related to the chromosomal <u>S</u>. <u>aureus</u> CadC (only 46% identical) and to an undefined open reading frame in the sequence of a different staphylococcal Cd^{2+} resistance system called CadB[1] (39% identities). CadC is also weakly homologous to the <u>arsR</u> gene product, which is the regulatory protein of the arsenic resistance system.[1] The resistance and transport data[31,34] indicate that CadC is an essential part of the CadA efflux complex, but how it functions is not known. The <u>Bacillus</u> CadC functions in <u>E</u>. <u>coli</u> to affect N^+/H^+ exchange,[33] and there is no current way to rationalize these diverse findings. However, the occurrence of additional polypeptides of unknown roles has happened also with other P-class ATPases. The enteric bacterial Kdp K^+- and MgtB Mg^{2+}-ATPases have additional polypeptides called KdpA and KdpC, or MgtC, respectively. The 122 amino acid sequence of CadC indicates that it is a soluble polypeptide, with 5 cysteine residues. There are also 4 histidine, 5 arginine, 12 lysine, 8 aspartate, and 9 glutamate residues, so that approximately 30% of the amino acids are charged, which is an unusually high value.

Transport inhibition studies suggested that the CadA Cd^{2+} efflux system was energy-coupled to the transmembrane proton gradient, as an electroneutral 2 H^+/1 Cd^{2+} exchange system.[30] Newer experiments of Cd^{2+} uptake via the CadA system in inside-out membrane vesicles of Bacillus cells show likely roles for both the proton gradient and for ATP,[31] supporting a more complex model in which both energy sources are required. Since the Na^+/K^+ system of animal cells is both an ATPase and a cation/cation exchanger, that may be the case with CadA as well.

The CadA system is inducibly activated, as demonstrated with gene fusions to a reporter gene[34] and by transport assays with subcellular membranes.[31] CadA is most sensitively regulated in response to environmental Cd^{2+}, and perhaps 50-times less effectively with Bi^{3+} and Pb^{2+}.[35] The regulatory gene for the CadA system (i.e. the hypothetical cadR gene), its location, and its mechanism have not been identified. However, such a regulatory gene must be located at a distance, since Cd^{2+} inducible regulation occurred with only a small 278 base pair fragment, which lacks any complete genes.[35] This short fragment includes only the start site for mRNA synthesis and the beginnning of the cadC gene[1,35] The mRNA start point for cadA is preceded by typical RNA polymerase binding motifs. The mRNA start site lies in the middle of a sequence of 7 bases plus 2 bases plus the inverted repeat of the 7 bases,[1,35] which is typical for binding sites of regulatory proteins. Partial removal of this inverted repeat eliminated gene regulation. These new and preliminary results for the CadA system open what will become a large study of gene regulation by divalent cations.

The cation regulation of the CadA system presents a pattern well-known to molecular geneticists, but less intuitive for inorganic chemists. This is the frequent finding that the substrate range for a biochemical process (such as a transport ATPase) may be quite different from that of its gene regulation. The reason is straight forward: different proteins and generally unrelated substrate binding sites are responsible for cation/protein recognition. The CadA system confers resistance most strongly to Cd^{2+} and considerably less effectively to Zn^{2+}. This may be because Zn^{2+} is a rather poor inducer of the cadA system.[35] However, Zn^{2+} resistance was readily measured after induced with Cd^{2+}, Co^{2+}, Pb^{2+}, and Bi^{3+}.[35] The CadA system does not confer resistances to Bi^{3+} and Pb^{2+}, which were effective inducers of Zn^{2+} resistance. Only once the CadR protein is isolated and direct cation/protein binding assays are possible, will these differences be explained.

4   NON-ATPASE Cd^{2+} EFFLUX SYSTEMS IN Alcaligenes

Alcaligenes eutrophus strain CH34 has separate plasmid-encoded efflux systems for Cd^{2+}, Zn^{2+} and Co^{2+} (the Czc system) and for Co^{2+} and Ni^{2+} (the Cnr system).[1,3] DNA sequences homologous to the czc operon are found elsewhere among bacteria from metal-polluted sites. The czc resistance system was initially shown to contain four genes in a single operon (Figure 6).[35] The functions of the first three gene products, CzcA, CzcB and CzcC were shown by deletion experiments to be involved in cation efflux. CzcA appears to be the central membrane transport protein for this system,[3,36] whereas CzcD appears to be involved in regulation.[3,36,37]

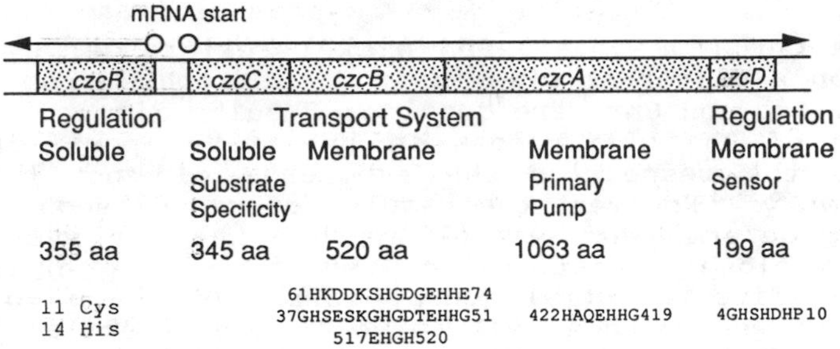

mRNA start

czcR	czcC	czcB	czcA	czcD
Regulation	Transport System			Regulation
Soluble	Soluble	Membrane	Membrane	Membrane
	Substrate Specificity		Primary Pump	Sensor
355 aa	345 aa	520 aa	1063 aa	199 aa
11 Cys 14 His	37GHSESKGHGDTEHHG51 517EHGH520	61HKDDKSHGDGEHHE74	422HAQEHHG419	4GHSHDHP10

Figure 6   Arrangment of genes in the Czc Cd^{2+}, Zn^{2+} and Co^{2+} efflux system.[3,35,36]

Deletion of the czcD gene had no effect on resistances to Cd^{2+}, Zn^{2+} and Co^{2+}, when mRNA was being controlled by an external regulatory system.[36] A partial deletion of the czcA gene eliminated efflux of and resistances to all three cations, Cd^{2+}, Zn^{2+} and Co^{2+}. On the other hand, a deletion involving only czcC resulted in sensitivity to (and failure to efflux) Cd^{2+} and Co^{2+}, but essentially full maintenance of resistance to and efflux of Zn^{2+}. Internal deletions in czcB produced strains that had residual Co^{2+} resistance but were sensitive to both Cd^{2+} and Zn^{2+}. The detailed correspondence of resistances and cation efflux rates resulted in a working model in which CzcA is the central protein of the Czc system and Zn^{2+} is the primary cation. CzcB and CzcC function as ancillary proteins expanding the range of cation specificities of the efflux system. The predicted CzcA, CzcB and CzcC protein sequences contain motifs rich in His residues, frequently associated with Gly, Glu and Asp residues. The tripeptide Glu-His-His is found both in CzcA and in CzcB (twice). His-Xaa-His sequences that might function as metal chelation sites were also found in CzcB (near the carboxyl end) and CzcD (twice near at the amino terminus).[3,36] Since other inorganic ion efflux systems are ATPases,[1] it is surprising that the Czc system shows no evidence of ATP involvement in cation efflux. The czc DNA sequence translated into proteins lacks recognizable ATP-binding motifs such as those found in known membrane ATPases.[3,35] Either a new pattern is yet to be determined or more likely the energetics of cation efflux via the Czc system does not involve ATP.

Regulation of the czc operon is being investigated. Growth resistance and gene fusions to reporter genes show that the system is inducible and responds to Cd^{2+}, Zn^{2+} and Co^{2+}. Two genes that act "in trans" (i.e. at a distance) have been identified.[3,37] czcD is postulated to encode a membrane protein that senses external cation concentrations (as internal concentrations are kept low by the Czc efflux pump). A new regulatory gene czcR was recently located and sequenced as part of the Czc system.[37] CzcR is required for cation-inducible function of the czc resistance and reporter gene fusions. An inverted symmetrical DNA sequence (as with cadA) is found between potential mRNA start sites for czcR and czcC. This is hypothesized to be the binding site for the trans-acting regulatory protein CzcR.[37]

The Cnr (for cobalt and nickel resistance) system is found on a different plasmid in A. eutrophus from the Czc system. Czc and Cnr were considered quite distinct systems because of the differences in substrate specificity and because the respective DNA fragments did not hybridize together.[36,38] However, the results of recent sequencing of the cnr determinant[39] establish that the two systems are actually closely related. The 8,500 base pair cnr sequence includes five potential genes. These include a long gene capable of encoding a 1076 amino acid polypeptide that is 46% identical to CzcA. Two shorter predicted gene products are 28% identical to CzcB and 30% identical to CzcC. The "downstream" cnr region comparable to czcD was not yet cloned or sequenced.[38] However, in the "upstream" region of cnr, comparable to where czcR is shown in Figure 6, the cnr sequence contains two possible genes that are thought to play regulatory roles. There is no currently-known similarity between these genes and czcR. Gene fusions with the cnr system respond strongly to Ni^{2+}, but only marginally to Co^{2+} (D. Nies, personal communication). The basis for this difference is yet to be determined but there is a new mutant class which adds Zn^{2+} resistance to the cnr determinant.[38] These Zn^{2+} resistant mutants also have increased Co^{2+} and Ni^{2+} resistance. Co^{2+}, Ni^{2+} and Zn^{2+} resistances in the mutant strains are constitutively expressed.[38] The Zn^{2+} resistant mutants therefore appear altered in gene regulation. For cnr as with czc, Zn^{2+} is a poor inducer of gene activity but Zn^{2+} still can be recognized by the membrane efflux system.

In summary, from the point of view of molecular geneticists and biochemists, bacterial cells have developed highly specific resistance mechanisms for divalent cations of five of the six elements in the Cu and Zn Triads. These mechanisms generally involve control of cellular cation transport. Our task for now is to define each of the components of these systems and their biochemical functions. Then we need to understand the rules for specificity of cation/protein recognition in both transport and regulatory proteins.

5   REFERENCES

1. S. Silver and M. Walderhaug, Gene regulation of plasmid- and chromosome-determined inorganic ion transport in bacteria. Microbiol. Rev. 56, 195-228, 1992.
2. N.L. Brown, D.A. Rouch and B.T.O. Lee, Copper resistance systems in bacteria. Plasmid 27, 41-51, 1992.
3. D.H. Nies, Resistance to cadmium, cobalt, zinc and nickel in microbes. Plasmid 27, 17-28, 1992.
4. N.L. Brown, J. Camakaris, B.T.O. Lee, T. Williams, A.P. Morby, J. Parkhill and D.A. Rouch, Bacterial resistances to mercury and copper. J. Cell. Biochem. 46, 106-114, 1991.
5. D.A. Cooksey, Copper uptake and resistance in bacteria. Molec. Microbiol., in press, 1992..
6. S. Silver and T.K. Misra, What DNA sequence analysis tells us about toxic heavy metal resistances, in "The Biological Alkylation of Heavy Elements," P.J. Craig and F. Glockling (eds.), Royal Society of Chemistry, London, pp. 211-242, 1988.
7. T.K. Misra, Bacterial resistances to inorganic mercury salts and organomercurials. Plasmid 27, 4-16, 1992.
8. C.T. Walsh, M.D. Distefano, M.J. Moore, L.M. Shewchuk and G.L. Verdine, Molecular basis of bacterial resistance to organomercurial and inorganic mercuric salts. FASEB J. 2, 124-130, 1988.
9. M.A. Mellano and D.A. Cooksey, Nucleotide sequence and organization of copper resistance genes from Pseudomonas syringae pv. tomato. J. Bacteriol. 170, 2879-2883, 1988.
10. J.-S. Cha and D.A. Cooksey, Copper resistance in Pseudomonas syringae mediated by periplasmic and outer membrane proteins. Proc. Natl. Acad. Sci. USA 88, 8915-8919, 1991.
11. D.A. Cooksey, Genetics of bactericide resistance in plant pathogenic bacteria. Annu. Rev. Phytopathol. 28, 201-219, 1990.
12. D.A. Cooksey, H.R. Azad, J.-S. Cha and C.-K. Lim, Copper resistance gene homologs in pathogenic and saprophytic bacterial species from tomato. Appl. Environ. Microbiol. 56, 431-435, 1990.
13. C. Ouzounis and C. Sander, A structure-derived sequence pattern for the detection of type I copper binding domains in distantly related proteins. FEBS _ Letters 279, 73-78, 1991.
14. H.S. Lee, A.H.T. Abdelal, M.A. Clark, and J.L. Ingraham, Molecular characterization of nosA, a Pseudomonas stutzeri gene encoding an outer membrane protein required to make copper-containing N_2O reductase. J. Bacteriol. 173, 5406-5413, 1991.
15. H. Cuypers and W.G. Zumpf, Regulatory components of the denitrification gene cluster of Pseudomonas stutzeri, in "Pseudomonas: Molecular biology and Biotechnology," E. Galli, S. Silver and B. Witholt (eds.), American Society for Microbiology, Washington, D.C., pp. 188-197, 1992.

16. M.A. Mellano and D.A. Cooksey, Induction of the copper resistance operon from Pseudomonas syringae. J. Bacteriol. 170, 4399-4401, 1988.

17. B.T.O. Lee, N.L. Brown, S. Rogers, A. Bergemann, J. Camakaris and D.A. Rouch, Bacterial response to copper in the environment: Copper resistance in Escherichia coli as a model system, in "Metal speciation in the environment," J.A.C. Broekaert, S. Gucer and F. Adams (eds.), NATO ASI Series Vol. G23, Springer Verlag, Berlin, pp. 625-632, 1990.

18. D.A. Rouch, Plasmid-mediated copper resistance in E. coli. Ph.D. thesis, University of Melbourne, 1986.

19. D. Rouch, B.T.O. Lee and J. Camakaris, Genetic and molecular basis of copper resistance in Escherichia coli, in "Metal ion homeostasis: Molecular Biology and Chemistry," D.H. Hamer and D.R. Winge (eds.), Alan R. Liss Inc., New York, pp. 439-446, 1989.

20. J.R. Williams, A. Morgan, D.A. Rouch, N.L. Brown and B.T.O. Lee, Copper resistant enteric bacteria from United Kingdom and Australian piggeries. Appl. Environ. Microbiol., submitted, 1992.

21. J.S. Parkinson and E.C. Kofoid, Communication modules in bacterial signaling proteins. Annu. Rev. Genet. 26, in press, 1992.

22. H.C. Tseng and C.W. Chen, A cloned ompR-like gene of Streptomyces lividans 66 suppresses defective melC1, a putative copper-transfer gene. Molec. Microbiol. 5, 1187-1196, 1991.

23. D. Rouch, J. Camakaris and B.T.O. Lee, Copper transport in Escherichia coli, in "Metal Ion Homeostasis: Molecular Biology and Chemistry," D.H. Hamer and D.R. Winge (eds.), Alan R. Liss, New York, pp.469-477, 1989.

24. S.D. Rogers, M.R. Bhave, J.F.B. Mercer, J. Camakaris and B.T.O. Lee, Cloning and characterization of cutE, a gene involved in copper transport in Escherichia coli. J. Bacteriol. 173, 6742-6748, 1991.

25. S. Silver, G. Nucifora, L. Chu and T.K. Misra, Bacterial resistance ATPases: primary pumps for exporting toxic cations and anions. Trends Biochem. Sci. 14, 76-80, 1989.

26. A. Odermatt, H. Suter, R. Krapf and M. Solioz, An ATPase operon involved in copper resistance by Enterococcus hirae. Ann. N. Y. Acad. Sci., in press, 1992.

27. D. McLaggan, T.M. Logan, D.G. Lynn and W. Epstein, Involvement of τ-glutamyl peptides in osmoadaptation of Escherichia coli. J. Bacteriol. 172, 3631-3636, 1990.

28. F.K. Gleason, C.-J. Lim, M. Gerami-Nejad and J.A. Fuchs, Characterization of Escherichia coli thioredoxins with altered active site residues. Biochemistry 29, 3701-3709, 1990.

29. A. Holmgren, Thioredoxin and glutaredoxin systems. J. Biol. Chem. 264, 13963-13966, 1989.

30. Z. Tynecka, Z. Gos and J. Zajac, Energy-dependent efflux of cadmium coded by a plasmid resistance determinant in Staphylococcus aureus. J. Bacteriol. 147, 313-319, 1981.

31. K.-J. Tsai, K.P. Yoon and A.R. Lynn, ATP-dependent cadmium transport by the cadA cadmium resistance determinant in everted membrane vesicles of Bacillus subtilis. J. Bacteriol. 174, 116-121, 1992.

32. G. Nucifora, L. Chu, T.K. Misra and S. Silver, Cadmium resistance from Staphylococcus aureus plasmid pI258 cadA gene results from a cadmium-efflux ATPase. Proc. Natl. Acad. Sci. USA 86, 3544-3548, 1989.

33. D.M. Ivey, A.A. Guffanti, Z. Shen, N. Kudyan, and T.A. Krulwich, The CadC gene product of alkaliphilic Bacillus firmus OF4 partially restores Na$^+$ resistance to an Escherichia coli strain lacking an Na$^+$/H$^+$ antiporter (NhaA). J. Bacteriol. 174, 4878-4884, 1992.

34 K.P. Yoon and S. Silver, A second gene in the Staphylococcus aureus cadA cadmium resistance determinant of plasmid pI258. J. Bacteriol. 173, 7636-7642, 1991.

35. K.P. Yoon, T.K. Misra and S. Silver, Regulation of the cadA cadmium resistance determinant of Staphylococcus aureus plasmid pI258. J. Bacteriol. 173, 7643-7649, 1991.

36. D. H. Nies, A. Nies, L. Chu and S. Silver, Expression and nucleotide sequence of a plasmid-determined divalent cation efflux system from Alcaligenes eutrophus. Proc. Natl. Acad. Sci. USA 86, 7351-7355, 1989.

37. D.H. Nies, CzcR and CzcD gene products affect regulation of cadmium, cobalt, zinc and nickel resistance in Alcaligenes eutrophus CH34. J. Bacteriol., submitted, 1992.

38. J.-M. Collard, A. Provoost, S. Taghavi and M. Mergeay, A new zinc resistance in Alcaligenes eutrophus CH34 generated by mutations affecting the regulation of the cnr cobalt/nickel resistance system. J. Bacteriol., submitted, 1992.

39. H. Liesegang, K. Lemke, R.A. Siddiqui and H.-G. Schlegel, Nucleotide sequence and expression of the inducible nickel and cobalt resistance determinant from pMOL28 of Alcaligenes eutrophus CH34. J. Bacteriol., submitted, 1992.

Metallothionein Genes from *Synechococcus* PCC6301 and PCC7942

Nigel J. Robinson, James W. Huckle, Jennifer S. Turner, Amit Gupta, and Andrew P. Morby

DEPARTMENT OF BIOLOGICAL SCIENCES, UNIVERSITY OF DURHAM, DURHAM DH1 3LE, UK

INTRODUCTION

In eukaryotes, metallothioneins (MTs) are involved in cellular responses to elevated concentrations of certain trace metal ions including those in the Cu and Zn triads (refer to 1). Genes encoding equivalent proteins have not previously been isolated from prokaryotes, although reports have indicated the presence of MT-like metal-ligands in several species (cited in 2). There is only one published amino acid sequence of such a protein, purified from *Synechococcus* sp. (3). Based upon this sequence PCR primers were synthesised. The resulting PCR products were used as probes to identify increases in the abundance of the corresponding transcripts following exposure to specific metal ions (4) and to isolate the corresponding MT locus, designated *smt*. The structure, function and regulation of the *smt* locus is reviewed.

STRUCTURE OF THE METALLOTHIONEIN LOCUS

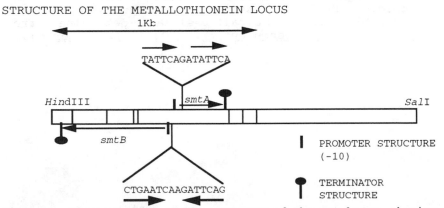

Figure 1 Representation of the structure of the *smt* locus, showing a 1.8 kb *Hind*III / *Sal*I fragment containing the MT gene, *smtA*, and divergent *smtB*. Sequence features within the *smt* operator-promoter region include a 7-2-7 hyphenated inverted repeat and a 6-2-6 hyphenated direct repeat located adjacent to, and 3' of, the *smtA* transcription start site. Vertical lines represent HIP1 sequences.

The MT locus from *Synechococcus* PCC7942 includes *smtA*, which encodes a class II MT, and a divergent gene *smtB* (Figure 1). The predicted sequence of the SmtA protein is similar to that purified from a *Synechoccocus* sp. (2) but with two additional amino acids at the C-terminus, N-His-Gly-C, and the substitution of Ser for Cys-32. The divergent ATG (encoding UAC translational start sites) sequences of both genes are separated by 100 bp. The sites of transcription initiation of both genes have been mapped within this region. Identified -10 consensus promoter motifs flank a 7-2-7 hyphenated inverted repeat (5'CTGAATC-AA-GATTCAG 3'). A 6-2-6 hyphenated direct repeat (5'TATTCA-GA-TATTCA 3') also lies within the region encoding the 3' untranslated portion of the *smtA* transcript.

METAL-BINDING PROPERTIES

To examine the metal-binding properties of its product the *smtA* gene was expressed in *E. coli* to generate SmtA as a carboxyterminal extension of glutathione-S-transferase (5). The protein was purified from *E. coli*, by glutathione affinity chromatography, and its amino acid sequence determined to confirm its identity. The pH of half dissociation of Zn, Cd and Cu ions from the expressed protein was determined to be 4.10, 3.50 and 2.35 respectively, indicating a high affinity for these metals (in particular for Zn in comparison to mammalian MT). Zn was displaced by Hg *in vitro*, using Zn-associated protein prepared from *E. coli* cells exposed to elevated concentrations of Zn for *in vitro* metal exchange. Associated Hg could not be displaced at low pH indicative of a high affinity for this metal.

CHARACTERISATION OF METALLOTHIONEIN DEFICIENT MUTANTS

Mutants deficient in functional *smt* have been generated via homologous-recombination-mediated insertional inactivation (6). A chloramphenicol acetyl transferase gene was inserted into the *smt* locus with the concomitant deletion of 373 bp including the *smt* operator-promoter region. Northern blots confirmed that the resultant *smt* cells were deficient in expression of *smtA* transcripts. Cultures of *smt* cells were hypersensitive to Zn and also showed reduced tolerance to Cd. No significant reduction in Cu tolerance was detected. *smt*-mediated restoration of Zn tolerance has subsequently been used as a selectable marker for transformation of these mutants.

METAL INDUCTION

smtA transcripts increase in abundance following exposure to elevated concentrations of certain trace metal ions including all of the Zn triad (Zn, Cd and Hg) and also Cu. Heat shock does not elicit an equivalent response indicating metal specificity, rather than a general "stress" response. Metal-induction is repressed by a transcriptional inhibitor (rifampicin) (6). There is no detectable effect of metal ions on *smtA* transcript stability with equivalent rates of transcript decay in rifampicin treated cells regardless of metal exposure.

Sequences upstream of *smtA* fused to a promoterless *lacZ* gene confer metal-dependent *beta*-galactosidase activity. At maximum permissive concentrations Zn is the most potent elicitor followed by Cu and Cd. At maximum permissive concentrations, Hg did not confer significant increases in *beta*-galactosidase activity although induction was observed in response to 2 h exposures to higher concentrations of Hg. Induction curves, showing steady state expression of *beta*-galactosidase at different concentrations of metal ions, indicate that the *smtA* promoter is ultrasensitive with respect to Zn and Cd.

The deduced SmtB polypeptide has structural similarity to ArsR (a metal-dependent repressor) and CadC proteins involved in resistance to arsenate/ arsenite/ antimonite and to Cd respectively. SmtB also contains a predicted helix-turn-helix DNA-binding motif. In *smt* mutants (also deficient in functional *smtB*) there is some residual induction of *beta*-galactosidase activity from the *smt* operator-promoter in response to Zn (possibly due to a general stimulation of rates of transcription by elevated Zn). Most significantly there is a substantial (>20 fold) increase in basal expression in non-metal supplemented media which declines (although not to normal basal levels, possibly due to the absence of SmtA) upon reintroduction of a plasmid borne copy of *smtB* (6). It is therefore proposed that SmtB is a repressor of *smt* transcription. Electrophoretic mobility shift assays (EMSA) have identified several protein complexes which associate with the *smt* operator-promoter region. Protein-binding sites include the 7-2-7 hyphenated inverted repeat and (a) site(s) located in a short region (39 bp) between the 7-2-7 hyphenated inverted repeat and the Shine/Dalgarno sequence of *smtA*. This region includes a 6-2-6 hyphenated direct repeat. Complexes which form with the latter region are absent in *smt* mutants, but are restored in *smt* mutants complemented with a plasmid borne copy of *smtB*. These data indicate that SmtB binds to a *cis*-element within this 39 bp region (6).

AMPLIFICATION AND REARRANGEMENT IN CADMIUM TOLERANT CELLS

Amplification and rearrangement of the MT locus has been observed in cells selected for tolerance to Cd (7). The rearranged locus has been cloned and sequenced from a Cd-tolerant cell line of *Synechococcus* PCC6301, C3.2. A 352 bp region of *smtB*, encoding the C-terminal portion of SmtB, is deleted in these cells. The functional deletion of *smtB* in cells selected for Cd-tolerance is consistent with the proposal that SmtB is a transcriptional repressor of *smtA* as derepressed expression of *smtA* may be beneficial for continuously metal challenged cells.

An octameric palindrome (5'GCGATCGC 3') traverses the borders of the excised element (6). Database analyses reveal that this is a highly iterated palindrome (HIP1) in the genomes of all cyanobacteria represented within the database. It occurs once every 664 bp in *Synechococcus* genera, and is present both within genes and in intragenic regions. HIP1 is therefore proposed to have a fundamental role in genome plasticity and hence adaptation to environmental change in these organisms.

ACKNOWLEDGEMENTS

This work is supported by research grant GR3/7883 from the Natural Environment Research Council. J.S.T and J.W.H. are supported by research studentships from the Natural Environment Research Council and A.G. is supported by a Jawaharlal Nehru Memorial Trust Scholarship. The authors thank J. Gilroy and J. Bryden for technical assistance and acknowledge the work of R.W. Olafson and co-workers. N.J.R. is a Royal Society University Research Fellow.

REFERENCES

1. J.F. Riordan and B.L. Vallee, Methods in Enzymology, 205, Academic Press, Inc., 1991.
2. S. Silver and T.K. Misra, Annu. Rev. Microbiol., 1988, 42, 717.
3. R.W. Olafson, W.D. McCubbin and C.M. Kay, Biochem J., 1988, 251, 691.
4. N.J. Robinson, A. Gupta, A.P. Fordham-Skelton, R.R.D. Croy, B.A. Whitton and J.W. Huckle, Proc. R. Soc. Lond. B., 1990, 242, 241.
5. J. Shi, W.P. Lindsay, J.W. Huckle, A.P. Morby and N.J. Robinson, FEBS, 1992, 303, 159.
6. Manuscript in preparation.
7. A. Gupta, B.A. Whitton, A.P. Morby, J.W. Huckle and N.J. Robinson, Proc. R. Soc. Lond B., 1992, 248, 273.

Chemical and Physical Characterisation of Zinc-replete Biocomposites

Julian G. McClements, Simon A. Smith, and Paul Wyeth*
APPLIED BIOCOMPOSITES GROUP, DEPARTMENT OF CHEMISTRY,
UNIVERSITY OF SOUTHAMPTON, SOUTHAMPTON SO9 5NH, UK

1 INTRODUCTION

The mouthparts of a variety of invertebrates have been shown to contain significant amounts of zinc[1] which seems to play an unfamiliar structural role. The metal is localised to the cutting edges where the dark, tanned cuticle is substantially harder than other regions[2]. More recently, zinc has also been detected in other structural appendages such as claws and stingers[3].

This study aimed to establish the total zinc content of the mandibular cuticle of the adult locust, *Schistocerca gregaria,* and confirm the levels in jaws of the ragworm, *Nereis virens*. The distribution of zinc in mouthpart cross-sections, of both species, was subsequently mapped and, in order to correlate the elemental composition and mechanical properties of metal-replete cuticle, hardness data were obtained on these same cross-sections.

2 TOTAL ZINC

Left and right mandibles of the locust *S. gregaria* were dissimilar, the left jaw being consistently the heavier of the pair. Although atomic absorption spectrophotometry showed that the absolute zinc content of the left mandible was also the higher, the level increased in proportion to the jaw weight as for the right mandible. When expressed as weight by dry weight of mandible the metal content was therefore the same for both (Table 1). Opposing jaws of the polychaete *N. virens* were of similar weight and showed no significant difference in the total metal content (Table 1).

Table 1 Total zinc content of the jaws of *Schistocerca gregaria* and *Nereis virens* as determined by atomic absorption spectrophotometry. The values are means of n samples, with standard deviation indicated.

Jaw Specimen (n)		Jaw Dry Wt (mg)	Zinc Content (μg)	Zinc Wt (%)
S. gregaria	Left(18)	3.7$\pm$0.1	9.3$\pm$0.4	0.25$\pm$0.01
	Right(18)	3.1$\pm$0.1	7.9$\pm$0.1	0.26$\pm$0.01
N. virens	Left (11)	0.62$\pm$0.07	10.8$\pm$0.9	1.79$\pm$0.09
	Right(11)	0.67$\pm$0.08	11.0$\pm$1.3	1.80$\pm$0.08

3 ZINC DISTRIBUTION

Mandibles of *Schistocerca gregaria*

Jaw cross-sections were viewed in a JEOL JSM 6400 scanning electron microscope. Associated X-ray microanalysis revealed an asymmetric distribution of zinc in the incisor regions of the locust mandibles. Figure 1b highlights the metal-replete zone in a longitudinal cross-section through a right mandible (Fig 1a). Zinc was localised to the inner face of the incisor cusps. X-ray mapping of left mandibles, however, demonstrated that the zinc was localised to the outer cuticular shell of the incisor region. X-ray point spectra (Fig. 1c) indicated the presence of chlorine, and high levels were shown to map coincidentally with those of the metal. Comparison with standards suggested that in the localised regions of both mandibles, zinc was present in concentrations of up to 7% of dry mass with the atomic ratio of zinc to chlorine ranging from 0.5 to 2.3.

Micromechanical Correlation. Vicker's diamond-point indentation tests showed that the zinc-replete regions of each mandible cross-section were at least twice as hard as any other part of the cuticle (Table 2). The differential localisation of metal to the outer face of the left but to the inner face of the right mandible, and the associated increase in cuticle hardness, lends the locust incisors a smart self-sharpening mechanism. During the shearing of food the softer, zinc-free cuticle wears preferentially leaving a prominent cutting edge to the incisor[2].

Jaws of *Nereis virens*

The zinc-replete region of the polychaete jaw (Fig. 2a) was rather more extensive; the metal permeated not only the tooth but the complete transverse cross-section at the tip and as far as the 7th distal tooth (Fig. 2b). X-ray composition mapping, again, showed a similar distribution for chlorine and zinc. For cross-sections through the mid-toothed region, the zinc content was up to 9% dry mass and the atomic ratio of zinc to chlorine ranged from 1.4 to 2.4. X-ray spectra also showed significant concentrations of bromine and iodine (Fig. 2c), particularly within the outer 50μm of the cuticle, as previously reported by Bryan and Gibbs[4]. Manganese was detected in cross-sections through the base of the jaw, localised in a region proximal to that of the zinc.

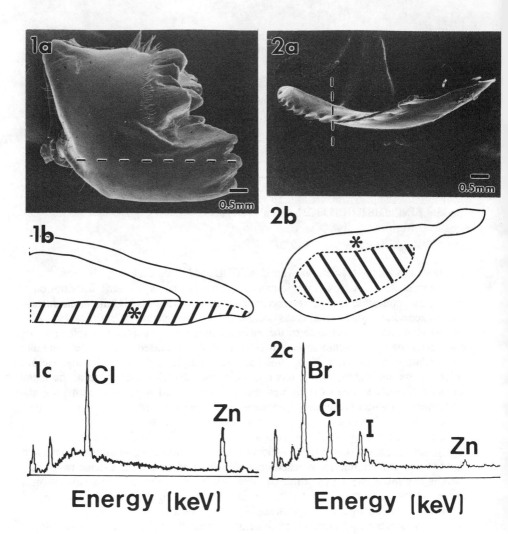

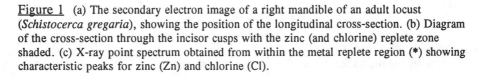

Figure 1 (a) The secondary electron image of a right mandible of an adult locust (*Schistocerca gregaria*), showing the position of the longitudinal cross-section. (b) Diagram of the cross-section through the incisor cusps with the zinc (and chlorine) replete zone shaded. (c) X-ray point spectrum obtained from within the metal replete region (*) showing characteristic peaks for zinc (Zn) and chlorine (Cl).

Figure 2 (a) The secondary electron image of a jaw of the marine polychaete *Nereis virens*, with the position of the transverse cross-section through the mid-toothed region shown. (b) Diagram of the whole section with the regions rich in zinc and chlorine shaded. (c) X-ray point spectrum from point * showing the presence of zinc (Zn), chlorine (Cl), bromine (Br) and iodine (I).

Table 2 Microhardness data obtained from cross-sections of mouthparts from *Schistocerca gregaria* and *Nereis virens*. The values are Vicker's hardness numbers (VHN) with standard deviation indicated.

Species	VHN from zinc-replete regions	VHN from zinc-free regions
S. gregaria	49.5±6.1	24.6±2.7
N. virens	64.4±7.6	44.0±6.6

Micromechanical Correlation. Microhardness tests showed that, where zinc was present, the hardness values were, on average, up to 50% higher than values obtained from metal-free jaw cross-sections (Table 2).

4   CONCLUSIONS

The jaws of both the locust, *Schistocerca gregaria*, and the annelid, *Nereis virens*, are loaded with high concentrations of zinc. As a percentage of total jaw dry weight, zinc content was invariant in each species; the value for the polychaete was higher reflecting the wider spread of the metal through the cuticle. But the values seem too low to implicate a high volume-fraction ceramic-filled composite, similar to that described for the radula teeth of the limpet, *Patella vulgata*[5]. Zinc may thus confer durability to invertebrate mouthparts in a novel way, perhaps providing high density bridging across organic polymers of the cuticle.

This parallel study on *S. gregaria* and *N. virens*, involving both element composition and micromechanical analysis, demonstrates that the presence of the zinc, directly or indirectly, is responsible for the hardening of the mouthpart working edges in these two species.

5   ACKNOWLEDGEMENTS

We must thank Julian Vincent (Centre for Biomimetics, Reading University) for introducing us to structural biomechanics. Financial support was given by the University of Southampton (SAS), SERC (JMC) and The Royal Society.

6   REFERENCES

1.   G.W. Bryan and P.E. Gibbs, J. mar biol. Ass. U.K., 1979, **59**, 969.
2.   J.E. Hillerton, S.E. Reynolds and J.F.V. Vincent, J. exp. Biol., 1982, **96**, 45.
3.   R.M.S. Schofield, PhD dissertation, University of Oregon, 1990.
4.   G.W. Bryan and P.E. Gibbs, J. mar. biol. Ass. U.K., 1980, **60**, 641.
5.   N.W. Runham, P.R Thornton, D.A. Shaw and R.C. Wayte, Z. Zellforsch, 1969, **99**, 641.

The Analysis and Speciation of Mercury Compounds in the Natural Environment

P. J. Craig, D. Mennie, and P. D. Bartlett

DEPARTMENT OF CHEMISTRY, DE MONTFORT UNIVERSITY,
THE GATEWAY, LEICESTER LE1 9BH, UK

1.   INTRODUCTION

General

The nature of mercury environmental pollution has changed considerably in the past twenty years.   The original problem arose in two different ways.   The first was in the deposition of considerable quantities of mercury as industrial effluent from the chloralkali industry (Hg^0), or from product use in agriculture (organic mercury). Industrial mercury entered water courses or lakes in tonnage quantities and was largely attached to bottom sediments.   From the use of organic mercury compounds in agriculture (viz CH_3HgCl, $CH_3OCH_2CH_2HgCl$ and $C_2H_5OCH_2CH_2HgCl$) organic mercury also entered biota.   Both of these problems have now been extensively reduced or eliminated.   In addition of course there were several large mercury poisoning incidents[1].

However, the problem still exists in more sophisticated form. For example, inorganic mercury enters the atmosphere through vulcanism, coal burning or incineration of household waste.   Although much point source discharge has been eliminated problems still arise.   For example high levels of mercury may be found in biota and sediments of lakes remote from any obvious source of mercury.   The deposition rate of mercury in southern Sweden has been estimated at $30g$ km^{-2} per year[2].   Despite measures of pollution abatement, methyl mercury levels in fish in some lakes subject to acid rain have increased in recent years[3]. This is due to increased supply of mercury due to acidification, and release of mercury (II) from atmosphere, humic binding etc.

It can be taken as a first approximation that nearly all the deposited mercury in a watercourse will be found in the sediments (>90%)[4]. It is here that the majority of the transformations take place. In this sense the sediments act as a holding reservoir for the mercury deposited and this may take many years to clear. Of the 150 tons of mercury emitted at Minamata about 45 tonnes have so far been removed to sea by natural causes. (A large scale decontamination project has also taken place at Minamata)[5].

It is well known that inorganic mercury entering the biosphere may be partially methylated to the more toxic methyl mercury or to dimethyl mercury. The former is more frequently found, but the latter may be the actual methylation product.[6] The proportion of methyl mercury to total mercury increases as the food chain is ascended. In sediments, methyl mercury may be in the range 0.1 - 8%, but is usually up to 1 - 2%.[7] Total mercury in sediments is probably converted to methyl mercury at the rate of 0.1% per annum.[8] In fish, nearly all the mercury is analyzed as CH_3Hg^+. The practical import of the environmental role of mercury as it relates to man is the concentration of methyl mercury in fish used for food. This may reach the ppm level. The normal level acceptable for health is 0.5 ppm.[9]

Clearly in view of the very different toxicities for inorganic and methyl mercury, analytical methods that discriminate a small amount of the latter in a larger amount of the former, needed to be found. Methyl mercury is toxic because the hydrophobic methyl group allows better transport to central nervous system cells compared to the more hydrophilic mercury (II) ion. The half life of CH_3HgCl in man is 72 days; that of $HgCl_2$ is 4 days.[10]

The Methylation of Mercury

Numerous substances, which may potentially be present in the environment, have been shown to convert abiotically a proportion of inorganic mercury present (Hg(0) or Hg (II)) in a system to methyl mercury. These include methyl cobalamin (CH_3CoB_{12}) (carbanion transfer, CH_3), other organometallic substances (eg $(CH_3)_3Sn^+$, $(CH_3)_3Pb^+$ etc), iodomethane (CH_3I) and other naturally occurring methyl species (via carbonium transfer, CH_3^+).[11,12]

Interestingly S Adenosyl methionine, the main biochemical provider of methyl carbonium ions, (CH_3^+) has not been reported to methylate inorganic mercury (Hg(II) or Hg(0)). It should also be observed that, whereas innumerable abiotic experiments show that CH_3CoB_{12} methylates mercury (II) it has not been demonstrated that environmental methyl mercury definitely arises from this species. Although experiments on tuna fish methylation provided circumstantial evidence for CH_3CoB_{12} involvement within the fish[13], current thinking is that inorganic mercury is methylated in the sediments and absorbed and concentrated by the fish. The reaction of CH_3CoB_{12} with mercury (II) is a carbanion transfer viz (Eqn 1).

$$CH_3CoB_{12} + Hg^{2+} \xrightarrow{\quad\quad} H_2O \longrightarrow CH_3Hg^+ + H_2CoB_{12}^+ \qquad [1]$$

The above is consistent with the fact that high salt concentration (with the formation of $HgCl_4^{2-}$) inhibits methylation.[4]

Oxidative addition of mercury (0) by, eg, CH_3I has been observed, but light is necessary (Eqn 2).[15]

$$CH_3I + Hg (0) \longrightarrow CH_3HgI \qquad [2]$$

Environmentally based work in sediments has not so far demonstrated that the above is a practical route. Such oxidative addition might be a general process for any carbonium ion doner reacting with mercury (0), but environmental evidence is lacking. Transalkylation has been proposed for many years as a source of alkylation and recent work has strongly argued the case (Eqn 3).[12]

$$(CH_3)_3M^+ + Hg^{2+} \longrightarrow (CH_3)_2M^{2+} + CH_3Hg^+ \quad M = Sn, Pb \qquad [3]$$

Many bacteria have been shown to be capable of both methylating and demethylating mercury.[16] In the environment the general characteristic for their success is low oxygen content (near anaerobic) and sulphate reduction ability in the bacteria. Lists are given in several sources, and include many of the common species (eg E Coli, S Brevicaulis etc). Numerous bacterial species also demethylate mercury.

Several general comments may be made about mercury methylation:

1. Mercury is methylated by organisms either as a detoxification process or adventitiously.

2. It can be methylated by both anaerobic and aerobic bacteria and by fungi, but formation of methyl mercury is higher the lower the oxygen concentration.

3. The rate of methylation is correlated with the general extent of microbiological activity.

4. Both methyl and dimethyl mercury may be formed, but at pH>8 it is mostly the latter.

5. Overall methylation rates are not much effected by pH variation between 5 and 9.

6. In the presence of sulphide ions, methyl mercury in sediments may be converted to dimethyl mercury which may volatilize to atmosphere and transportation.

7. Demethylation of methyl mercury occurs by bacteria, light etc.

8. Excess sulphide, producing HgS, inhabits methylation as mercury is not in a form available for methylation.

The role of sulphide may perhaps be the main controlling feature in mercury methylation.[17] At moderate levels of H_2S, dimethylmercury is produced (Eqn 4).

$$2CH_3Hg^+ + S^{2-} \longrightarrow (CH_3Hg)_2 S \longrightarrow$$
$$(CH_3)_2Hg + HgS \quad [4]$$

At higher levels of sulphide, the substrate mercury (II),
is converted to HgS and no methyl mercury is present - any
mercury present as dimethyl mercury having diffused to
atmosphere being volatile and hydrophobic. It has been
calculated that up to 13% annually of methyl mercury
present may be converted and removed as dimethyl mercury by
this route.[18]

It is also apparent that inorganic mercury speciation as
well as concentration is important. We note that species
containing a Hg-Cl bond have an inhibiting effect on
methylation in sediments comparing with Hg-S linkages in
aminoacid complexes. This is related to the ability of
microorganisms to break down the mercury complexes into a
form (mercury (0) or (II)) suitable for methylation.[19] In
the environment anionic mercury species (eg $HgCl_4^{2-}$ in
seawater) are less easily methylated.[14]

The original analytical methodologies for total and methyl
mercury in various environmental matrices were originated
more than 20 years ago. However, the classic methods
described in the next section are still at the core of the
methods most frequently used today, and they are indeed
used in our own Group. As such they are reviewed in a
following section. However, as will be seen, the
limitations in these methods are currently providing an
imperative for new and more up to date methods. The
earlier methods for methyl mercury will now be reviewed.
As more than 90% of mercury by mass will be present in
sediments, and as toxicologically important mercury exists
in biota, analysis from these complex matrices will be
emphasized.

2.   THE DETERMINATION OF TOTAL MERCURY IN SEDIMENT
     SAMPLES

Many analytical techniques capable of detecting trace
levels have been developed. The more important works, with
particular emphasis on sediment analysis are discussed in
Reference 1.

The majority of these techniques require a sample
preparation stage followed by analysis of the mercury
content.   This usually involves oxidative extraction,
followed by reduction using $SnCl_2$ and cold vapour detection
of mercury (0).   Space precludes a description of total
mercury analysis, but the main methods are reviewed in
Reference 1.

3 THE DETERMINATION OF METHYL MERCURY IN ENVIRONMENTAL
SEDIMENT SAMPLES
Extraction techniques

Gage[20] in 1961 produced the first practical method for
differentiating between inorganic and organic mercurial
compounds in tissues. The Ch_3HgCl was extracted into $CHCl_3$.
There it could be measured spectrophotometrically as the
dithizone complex.   Gage found that methyl and phenyl
mercury compounds could be extracted into C_6H_6 from tissue
homogenates in the presence of high concentrations of HCl.
Inorganic mercury remained in the aqueous phase.   The
organic mercurials were further extracted into aqueous Na_2S
and the mercury content determined by dithizone titration
following $KMnO4$ oxidation. Recoveries averaged 90% and the
detection limit was 1 µg g^{-1}. Magos[21] described a method of
differentiating between organic and inorganic mercury in
undigested biological samples. This was a reduction, cold-
vapour atomic absorption method. The organo-mercurials can
be rapidly reduced to mercury vapour by a $CdCl_2/SnCl_2$
mixture. $SnCl_2$ alone released only the inorganic mercury.
A specific analysis for methyl mercury first appeared in
1966 and was developed by both Japanese and Swedish

workers[22,23]. Both methods used the Gage procedure for the
extraction of CH_3HgCl into organic solvents after strong
halogen acid treatment. Determination of methyl mercury
halide was by gas chromatography (GC) using an electron
capture detection (ECD) system. These detectors are
extremely sensitive to materials that are 'electron-
capturing', eg halides. Kitamura[24] extracted CH_3HgCl from
the acidified sample into $CHCl_3$ and then into C_6H_6 for
chromatography. No detailed clean-up steps were employed
so that interferences on the chromatograph from both the
$CHCl_3$ and other extracted substances were present.

Westoo[22,23] employed a similar procedure, but with the use of
a double partitioning clean-up procedure to remove
interferences from fish extractions. The sample was
extracted with C_6H_6 and HCl, then a portion of the C_6H_6 was
partitioned with an aqueous solution of cysteine
$(HSCH_2CH(NH_2)COOH)$.

A procedure based on that of Westoo was adopted by the
Swedish Water and Air Pollution Research Laboratory as
their standard method for determination of methyl mercury
by gas chromatography. Here the clean-up procedure
partioned the methyl mercury into an aqueous phase and then
back as a sulphur complex into a clean organic solution.
Sulphur groups were masked with copper ions rather than
mercuric ions to avoid decomposition of any $(CH_3)_2Hg$
present. Methyl mercury was extracted as the bromide,
rather than the chloride, because the partition
coefficients are more favourable. The methyl mercury is
initially released from the substrate by copper and bromide
ions in strong H_2SO_4 and extracted into C_6H_6. It is then
back-extracted as a sulphur complex into dilute aqueous
$Na_2S_2O_3$ instead of cysteine. This is treated with strong
copper bromide following which the methyl mercury is re
extracted as the bromide into benzene for determination by
GC-ECD. Recoveries of 85-90% of added methyl mercury were
reported together with a detection limit of 0.5 ng g^{-1} from
1g of sample.

Uthe[25] reduced the basic procedure to a semi-micro scale
using special vials to perform the partitioning steps. The
initial C_6H_6 or $C_6H_5CH_3$ extraction of the fish sample is
performed by grinding in a ball mill. This is followed by
a clean-up procedure using ethanolic $Na_2S_2O_3$. The methyl
mercury was then re-extracted into C_6H_6 and determined as
the iodide complex. 99% recoveries were reported and few
interferences were noted. $(CH_3)_2Hg$ was decomposed by only
1-2% illustrating the advantages of copper over the use of
mercuric ion.

Method of Methyl Mercury Analysis Used

(i) Extraction Procedure

Methyl mercury in sediments is determined by the method of
Bartlett, Morton and Moreton[26]. This is a semi-micro
procedure developed from work by Longbottom[27] and Uthe[25].
To increase stability, Ch_3HgBr is now collected in the
final extract instead of the iodide.

Longbottom[68] had published one of the few methods
specifically for sediments. The sediment is extracted with
copper, bromide and H_2SO_4 into $C_6H_5CH_3$ in a stoppered $50cm^3$
centrifuge tube. The $C_6H_5CH_3$, after removal, was back
extracted into aqueous $Na_2S_2O_3$. The methyl mercury was re-
extracted with excess iodide into C_6H_6. Our method is based
on that.

(ii) Apparatus and Reagents

Cleanliness of glassware is of critical importance in this
procedure. For most of this study a 5% v/v solution of
Decon-90 detergent was used for cleaning and storage of
glassware. This was used instead of a chromic acid bath,
but it was necessary to ensure complete removal and rinsing
of the glassware. On occasions some difficulty was
experienced in the separation of toluene-thiosulphate
emulsions. This seemed to occur with specific types of
sediments, but some evidence connected this with glassware
bearing traces of detergent.

The method used for the analysis of methyl mercury in

sediments is detailed below:

(iii)   Method

2-5g of wet sediment is weighed into a 35cm^3 centrifuge
tube.   The moisture content of another portion of the
sample is determined by drying at 110^0.   The sample is made
up to a volume of 10 cm^3 with distilled water.   7.5 cm^3 of
toluene (A.R.) is added followed by 1 cm^3 of 25% W/V CuSO$_4$
and 4 cm^3 of acidic KBr reagent.   When the effervescence
reduces, the tube is securely stoppered and shaken
vigorously.   It is then centrifuged to separate the solid,
aqueous and organic phases.   Generally some material is
held at the interface between organic and aqueous phases.
If this is excessive, the interface is stirred gently with
a glass rod, and the mixture re-centrifuged, until at least
4 cm^3 of the toluene is clear.   A measured volume of
toluene is removed and transferred to a 10 cm^3 centrifuge
tube.   This is extracted twice with 3 cm^3 portions of the
0.005M Na$_2$S$_2$O$_3$ reagent; the aqueous layers are removed with
a Pasteur pipette, and combined in another centrifuge tube.
Complete recovery of the aqueous phase is normally
possible.   It is sometimes advantageous to centrifuge the
extraction mixtures to obtain complete separation.   The
thiosulphate extracts are treated with 0.2 cm^3 of the
copper reagent, 1 cm^3 of bromide reagent and 1 cm^3 of C$_6$H$_6$
or C$_6$H$_5$CH$_3$.   This mixture is shaken vigorously by hand for
two minutes, then the organic layer is removed by pipette
and stored over anhydrous Na$_2$SO$_4$ in small glass vials.
These are kept in the refrigerator (2^{0}C) until the
solutions can be analyzed by gas chromatography.   A reagent
blank is run in parallel with the above procedure.

An internal standard may be used to assist the calibration
of the chromatography.   In this case 1 cm^3 of C$_2$H$_5$HgCl in
C$_6$H$_6$ or C$_6$H$_5$CH$_3$ (100 or 20 µgdm^{-3}) can be added to the final
extract instead of solvent alone.

(iv) Gas Chromatography of Methyl Mercury

Various instruments have been used in our Group all using
an electron capture detector.   This detector uses a [63]Ni

source. High purity oxygen-free nitrogen is used as the
carrier gas. The use of a molecular sieve filter provides
a further purity safeguard. Carrier gas flow is typically
set at 100 cm^3 min^{-1}. The optimum instrument temperature
used are: detector 265^0, column 165^0, column inlet 110^0.
The column type used throughout this work is a glass
column, 0.4 cm diameter, with a Pye glass to metal seal at
the exit. The column length varied with the packing type,
but is 0.9m or 1.5m. The packing used is monoethylene
glycol adipate on Chromosorb G (AW) 80-100 mesh. This
column packing is conditioned for several days at 190^0 and
is then found to give a satisfactory performance on
standards and samples. When not in use the column
temperature is reduced to 110^0, the injection heater turned
off and the carrier-gas flow rate reduced to 60 cm^3 min^{-1}.
Performance of the column decreases slowly with usage.
Attempts at revitalisation meet with limited success.
Various authors have suggested devices to regain column
efficiency. These were found to be ineffective or caused
greater deterioration. Reconditioning the column at high
temperature, or re-packing the first few centimetres,
showed only marginal improvements. Complete replacement
with a fresh column is necessary. Columns are replaced
when performance declined. This is necessary about every
3 months, but clearly depends on the amount of usage. The
electron capture detector is run at high temperature
periodically to remove any deposited material. It is
necessary to have nitrogen carrier gas flowing at all times
when the detector is hot. The injection size normally used
was 1μL; however, injections up to 10μL were possible.
Injections of standards were performed at frequent
intervals during the analyses to assess performance. The
standard solution most commonly used contained 20 μg dm^{-3}
methyl mercuric chloride in benzene or toluene (Ultrar).

Use of calibration and standard solution was the basic
method of quantification with ethyl mercury used as an
internal standard. The practical detection limit in the
sedement is 0.3 ng g^{-1}. Recovery of added methyl mercury

was 95%. Several examples of published work in both environmental and natural matrices exist using this method.[17-19,28]

(v)   Limitations

The limitations of this method may be briefly discussed as follows:

A number of problems exist with the non specific methodology used in the analysis of methyl mercury by electron capture detection.[29]

1.   Formation of emulsions during the extraction of methyl mercury may occur, leading to losses.

2.   Distribution coefficients between aqueous and organic solvents in the extractions are not high (about 5 to 10) making it difficult to concentrate methyl mercury from natural waters by more than 5×10^3.

3.   Severe tailing on chromatography often occurs, and there is often decomposition on the column, whose efficiencies are often low. Conditioning of the column has to be carried out frequently. Capillary columns do not usually improve matters.

4.   The ECD is not compound or element specific and so chromatograms may be complex and subject to interferences.

5.   Non polar dialkyl derivatives of methyl mercury would be expected to give superior chromatographic characteristics, but electron capture detection is not so sensitive here and mercury specific detectors are required.

The latter has been the thrust of the recent analytical developments for methyl mercury. These new methods have all essentially aimed at the conversion of methyl mercury to a dialkyl form and identification by more sensitive and specific means, usually GC interfaced to atomic fluorescence (AF), atomic absorption (AA) or mass spectroscopic (MS) detectors.

NEW DEVELOPMENTS IN THE ANALYSIS OF METHYL MERCURY

Bulska et al have butylated monoalkyl mercury species using a Grignard reagent. Following separation by gas chromatography, identification was by microwave induced plasma emission detection.[30]

The use of $NaB(C_2H_5)_4$ for the liberation of organometallic species from biota and sediments, and for subsequent gas chromatographic separation of the fully saturated alkyl (eg $(C_4H_9)_3SnC_2H_5$ has been reported in recent years. Detection and identification is usually by AA or MS interfaced with the GC.[31]

Bloom[32] has recently described in detail the use of $NaB(C_2H_5)_4$ in derivatisation of mercury compounds. He used $NaB(C_2H_5)_4$ to produce $Hg(C_2H_5)_2$ from labile mercury (II) compounds and $CH_3HgC_2H_5$ from methyl mercury. The volatile ethyl derivatives were collected on a cryogenic column and subsequently detected by a cold vapour atomic fluorescence method. The method was used for mercury in water samples at the pg-ng L^{-1} level, and for mercury levels in fish at the ng g^{-1} level.

We have recently carried out further work on this system[33], also using a capillary column giving full chromatographic integrity from injection point to detection cell.[34] Using the above methodology an aqueous solution containing a single mercury compound produces only a single AA peak on ethylation confirming that no dismutation was taking place. A solution in water containing $HgCl_2$ and CH_3HgCl (or ones containing CH_3HgCl and C_2H_5HgCl respectively) in water each produced two large peaks on ethylation whose peak area was in proportion to the original concentrations with little evidence of dismutation. It was necessary for the $NaB(C_2H_5)_4$ solution to be in 4 fold molar excess over the mercury analyte. From solutions containing $10cm^3$ of 0.005-1ppm mercury (II) or methyl or ethylmercury chloride derivatized products were delivered to a headspace volume of $10cm^3$. From this, $1cm^3$ of headspace mercury was analysed by injection (ie 5ng-1µg range). This methodology is therefore convenient for aqueous solutions of mercury compounds compared with other interfaced techniques.

Extraction of methyl mercury from $C_6H_5CH_3$ solutions was carried out as follows. $10cm^3$ of $C_6H_5CH_3$ was extracted with freshly prepared $5cm^3$ aliquots of 0.005M $Na_2S_2O_3$ solution with a magnetic stirrer for 10 minutes each. Separation of the organic/aqueous phase was achieved via phase separating filter papers (Whatman) and the two aqueous extracts were combined in a $20cm^3$ crimp top vial prior to analysis as above. Detection limit was 0.01 µg g^{-1} (ppm) from the $C_6H_5CH_3$ solutions, compared to a simple aqueous solution detection limit of 0.005 µg g^{-1} implying incomplete generation of the mercury ethyl from this solution.

This analysis was also carried out on 5g of a dried and powdered tuna fish muscle. The sample was spiked with $5cm^3$ of 1µg g^{-1} CH_3HgCl aqueous solution and left overnight. It was then extracted by the following method, the tuna fish sample being initially at a nominal 1µg g^{-1} in CH_3HgCl.

Two grams of the tuna fish muscle were weighed into a $100cm^3$ beaker and suspended in $15cm^3$ of 10% H_2SO_4 and left overnight. This was then extracted for ten minutes using a magnetic stirrer and follower. The solid material was then removed via vacuum filtration. Next $10cm^3$ of $C_6H_5CH_3$ was used to extract the aqueous phase. Phase separating filter papers were used to separate the organic from the aqueous phase and were found to give excellent results, and were able to separate the organic/aqueous phase emulsions that the stirring generates. The $C_6H_5CH_3$ extract was then extracted using two $7.5cm^3$ aliquots of freshly prepared 0.005M $Na_2S_2O_3$ solution. Phase separating filter papers were again used to separate the organic/aqueous phase emulsions.

The $Na_2S_2O_3$ extracts ($15cm^3$) now containing methyl mercury were combined in a $20cm^3$ crimp top vial which was sealed with a PTFE-faced butyl rubber septum. The extract was derivatized by needle injection using $1cm^3$ of a 1% aqueous solution of $NaB(C_2H_5)_4$. This was left for 30 minutes prior to injection of $1cm^3$ of the headspace to the GC AA system. Derivatization from the thiosulphate extracts was equal in yield to derivatization from standard thiosulphate solutions. However, we did find that yields from thiosulphate containing aqueous solutions are only 50% of those from aqueous solutions not containing thosulphate. This is allowed for in the calculation.

By these means fish samples containing either 0.1ppm or 1ppm of methyl mercury were analyzed. The method is also successful for samples containing 10ppb using 10g of fish muscle, bringing the method within the range of mercury concentration limits for fish.

The analysis of mercury in sediment matrices was carried out as follows; initially 100g of sediment was doped with $1cm^3$ aliquots of ethanolic solutions of 1000ppm CH_3HgCl and/or $HgCl_2$. These were left for twenty four hours prior to extraction.

Five grams of sediment were then suspended in $15cm^3$ of 10% H_2SO_4, left overnight and stirred vigorously with a magnetic stirrer and follower for ten minutes. The mixture was then filtered using a phase separating filter paper.

The $C_6H_5CH_3$ extract which contained the mercury species was then extracted by stirring with a magnetic stirrer for ten minutes with two separate $9cm^3$ aliquots of freshly prepared 0.005M $Na_2S_2O_3$ solution. The $C_6H_5CH_3$ was separated from the $Na_2S_2O_3$ layer via a phase separating filter paper which again was found to give excellent separation avoiding the problems of emulsions which other separating methods tended to generate. The aqueous phases were then combined in a $20cm^3$ crimp top vial sealed with a PTFE faced butyl septum. $1cm^3$ of a 2% solution of $NaB(C_2H_5)_4$ was then injected into the vial which was left for 30 minutes prior to analysis. $1cm^3$ of the $10cm^3$ headspace was then injected in the GC AA for analysis.

This method has so far proved successful on sediments down to the ng g^{-1} level frequently observed in the natural environment.

The analysis from sediment matrices is simpler than previous methods and it is element specific. GC ECD traces in particular require close attention to the identification of the mercury analyte amidst several or many other peaks if the matrix is complex.

We have observed very little evidence of dismutation of the derivatized samples viz, CH_3HgCl when derivatized produced only $CH_3HgC_2H_5$. Identification of peaks was also confirmed by mass spectra.

That there is a lack of dependence on hydrogen in the transfer gas suggests that the analytical process in the quartz cell depends only on the thermal degradation to mercury atoms of the alkyl mercury derivative and that it does not depend on a hydrogen radical induced decomposition. This is further substantiated by the dimethyl mercury species showing a greater peak area at 200^0C then the methyl ethyl species. This is assumed to be due to the relative bond strengths of the two species with the ethyl group being removed more easily allowing a higher proportion of ground state mercury atoms to be presented in the AA light path. The hydrogen flow appears to reduce analytical sensitivity by reducing the residence time of the mercury in the cell (ie dilution).

Similar derivitization and detection using MS confirms the separate or mixed generation of $CH_3HgC_2H_5$ and $(C_2H_5)_2$ Hg without dismutation.

Use of NaBH$_4$

Sodium Borohydride (NaBH$_4$) is frequently used to generate stable and volatile organometallic hydrides for analysis, eg $(C_4H_9)_3SnH$ from $(C_4H_9)_3Sn^+$ in various environmental matrices. However, the reaction of CH_3HgCl with $NaBH_4$ is thought to proceed as follows (Eqn 5)[35]

$$4RHgCl + NaBH_4 + 4OH^- \longrightarrow$$
$$4RH + 4Hg + 4Cl^- + Na^+ + H_2BO_3 + H_2O \qquad [5]$$

There is now evidence, however, that CH_3HgH can be generated from aqueous solution by $NaBH_4$ and can be purged with helium and condensed in a cold trap and distilled for AA detection or it can be transported directly after generation along a capillary gas chromatography column and then be subjected to AA or MS detection. We recently reported direct MS and NMR evidence for the existence of CH_3HgH and CH_3HgD.

A literature search for CH_3HgH produced only a single reference quotation$_2$. In her Communication[36], Devaud presented evidence for the existence of RHgH species in methanol solution from their polarographic oxidation wave. She derived rate constants for the decomposition of the RHgH species. However recent work by Donard[37] et al (with AA detection) suggested the existence of CH_3HgH and Craig[38,39] et al have recently obtained MS evidence.

For the detection the usual forms of apparatus have been used. One is a capillary GC system linked by a transfer line to an AA instrument used as detector. AA detection is via a quartz cell at 600^0. Alternatively a commercially interfaced GC MS may be used.[38,39] CH_3HgH has also been obtained by a purge and trap system.[37] The volatilized CH_3HgH is trapped at liquid nitrogen temperature in a glass U-tube. The trap is electrothermally heated and the CH_3HgH volatilizes and is detected by atomization in a quartz cell in the beam of an AA spectrometer. Purge and trap derivatization was carried out on solutions at around the 50ng CH_3HgCl per $50cm^3$ H_2O level (1ppb), in this system the recovery of CH_3HgH from CH_3HgCl was 60%.[37] Purge and trap AA, or GC AA methodologies, allow partial specification of the substance being analyzed to the extent of elemental confirmation and retention of CH_3HgCl at the 1000ppb level.

However, only MS data can establish full speciation.

The clearest evidence for the existence of CH_3HgH comes from the Mass Spectra (MS).[38] The MS of the proposed CH_3HgH shows no peaks centred on m/e 252 or 232 ie it is not a spectrum of underivatized CH_3HgCl or $(CH_3)_2Hg$. A spectrum of CH_3HgCl or $(CH_3)_2Hg$ shows the expected CH_3Hg^+ pattern centred on m/e 217, viz the expected mercury isotopes in CH_3Hg^+ with a maximum value of m/e 219 for CH_3Hg^+. CH_3HgCl derivatized with $NaBH_4$ shows peaks we assign to CH_3HgH clustered around m/e 216 and 218 consisting of CH_3Hg^+ and CH_3HgH, with a maximum value of m/e 220 for CH_3HgH. CH_3HgCl derivatized with $NaBD_4$ shows peaks clustered around m/e 217 and 219 with a maximum value at m/e 221 for CH_3HgD. The same follows for the $^{204}Hg^+$, $^{204}HgH^+$ and a ^{204}HgD peaks in the MS at m/e 204, 205 and 206 respectively. Appropriate mixtures are found for $^{204}Hg^+$ with $^{204}HgH^+$ and $^{204}HgD^+$ respectively.[38,39]

These arguements have also been supported by recent NMR and also by independent work elsewhere. NMR investigation has shown the existence of a methyl group connected to mercury that exists as a doublet owing to coupling with the mercury - hydrogen atum (which itself is a quartet). These couplings were established by use of spin decoupling techniques.[38]

Conclusion

It can be seen that in principle the newer, speciation methods for methyl mercury should offer a cleaner and more certain methodology for analysis, and no doubt they will make increasing headway in environmental analysis in future years. However, it is still probably true that the majority of fieldwork investigations for methyl mercury still use variants of the GC-ECD technique.[41]

References

1. Craig, P.J. (ed), Organometallic Compounds in the Environment, Longman, 1986.
2. Lindqvist, O, and Schroeder, WH in Pacyna, J.M. and Ottar, B(eds), Control and Fate of Atmospheric Trace Metals, Kluwer, 1986.
3. Anderson, I., Parkman, H., and Jemelov. A, Limnologica, 1990, 20, 347.
4. Kudo, A., Miller, D.R., Akagi, H., Mortimer, D.C., de Freitas, A.S., Nagase, H. Townsend, D.R., and Warnock, R.G., Prog. Water. Technol., 1978, 10, 329.
5. Kudo, A., and Miyahara, S., Wat. Sci. Technol., 1991, 23, 283.
6. Quevauviller, Ph., Donard, OFX, Wasserman, J.C., Martin, F.M. and Scheider, Appl. Organometallic Chem, 1992 (in press).
7. Ref. 1, p 92.
8. Ref. 2, p 351.
9. Varta, M., Can. J. Fish. Aquat. Sci., 1990, 47, 1888.
10. Ref. 1, p 72.
11. Ref. 1, p 86.
12. Ebinghaus, R. and Wilken, R.D., Appl. Organometallic Chem., , 1992 (in press).
13. Imura, N., Pan, S-K., Shimitzu, M., Ukita, T., 1973 In New Methods Environ. Chem. Toxicol., Collected Papers Res. Conf. New Methods Ecol. Chem. (Coulston, F., Ed) Int. Acad. Print. Co., Tokyo, 211.
14. Compeau, G., and Bartha, R., Appl. Environ. Microbiol., 1984, 48, 1203.
15. Maynard, J.L., J. Amer. Chem. Soc., 1932, 54, 2108.
16. Ref. 1, p 87.
17. Craig, P.J., and Bartlett, P.D., Nature, 1978 275 635.
18. Craig, P.J., and Moreton, P.A., Mar. Poll. Bull. 1984, 15, 406.
19. Craig, P.J., and Moreton, P.A., Environ. Poll. Ser. B, 1985, 10, 141.
20. Gage, J.C., Analyst, 1961, 86, 457.

21.   Magos, L., Analyst, 1971, 96, 847.

22.   Westoo, G., Acta. Chem. Scand., 1966, 20 2131.

23.   Westoo, G., Acta. Chem. Scand., 1967, 21 1790.

24.   Kitamura, S., Tsukamoto., T., Hayakawa, K., Sumino, K., Shibata, T. Ig Seibutsugaku, 1966, 72, 274.

25.   Uthe, J.F., Soloman, J., and Grift, B., J. Assoc. Off. Anal. Chem., 1972, 55, 583.

26.   PhD Theses, Leicester Polytechnic, Leicester, UK, 1977, 1979, 1983.

27.   Longbottom, J.E., Dressman, R.C., Lichtenburg, J.E., J. Assoc., Off. Anal. Chem., 1973, 56, 1297.

28.   Bartlett, P.D., Craig, P.J., and Morton S.F. Sci. Tota. Environ. 1978, 10, 245.

29.   Wilken, R-D., Hintelmann, H., Nato ASI series Vol. G23 Metal Speciation in the Environment (Broekaert, J.A.C. Gucer, S., and Adams F. (eds.)), Springer, 1990, 339.

30.   Bulska, E., Baxter, D.C., and Frech, W., Anal. Chim. Acta, 1991, 249, 545.

31.   See for example Mason, R.P., and Fitzgerald, W.F., Nature, 1990, 347, 457.

32.   Bloom, N., Can. J. Fish. Aquat. Sci., 1989, 46, 1131.

33.   Rapsomanikis, S., and Craig, P.J., Anal. Chim. Acta, 1991, 248, 563.

34.   Craig, P.J., and Mennie, D. In preparation, 1992.

35.   Wardell, J.L., in Comprehensive Organometallic Chemistry (Wilkinson G, Store FGA and Abel EW (eds)) Pergamon Press 1982, vol 2, p 966.

36.   Devaud, M., J. Organometallic Chem., 1981, 220, C27.

37.   Donard, O.F.X., Sond. aus Colloq. Atomspek. Spurenanalytik., Bodenseewerk Perkin Elmer GmBH, D7770 Uberlingen, Germany, 1989.

38.   Craig, P.J., Mennie, D., Needham, M., Ostah, N., Donard, O.F.X., and Martin, F., J. Organometallic Chem., 1992 (accepted and in press).

39.   Craig, P.J., Mennie, D., Ostah, N., Donard, O.F.X., and Martin, F. Analyst 1992, 117, 823.

40.   Fillipelli, M., Baldi, F., Brinckman, F., and Olson, G.J., Environ. Sci. Technol., 1992, 26, 1457.

41.   Bubb, J.M., Rudd, T., and Lester, J.N., Sci. Total. Environ., 1991, <u>102</u>, 147.

Acknowledgement

The receipt of a studentship (DM) from the Science and Engineering Research Council UK is gratefully acknowledged. Receipt of funding from the British Council is also gratefully acknowledged.

An Investigation of the Zinc Sites of Hydrolytic Enzymes via a Model Compound Approach

A. K. Powell[1,*], A. C. Deveson[1,*], D. Collison[2], D. R. Harper[2], and F. E. Mabbs[2]

[1] SCHOOL OF CHEMICAL SCIENCES, UNIVERSITY OF EAST ANGLIA, NORWICH NR4 7TJ, UK

[2] DEPARTMENT OF CHEMISTRY, UNIVERSITY OF MANCHESTER, MANCHESTER M13 9PL, UK

Zinc fulfils both structural and catalytic roles in biology. We have been investigating the active sites of hydrolytic enzymes using a model compound approach. In these the Zn^{2+} can activate water and small substrates such as carbonate as in, for example, carbonic anhydrase.[1,2] Often the Zn^{2+} in such enzymes has been substituted with different M^{2+} ions which have useful spectroscopic properties enabling the metal site to be probed. In many cases this is accompanied by a loss of enzyme activity. The reasons for this are not always clear, but may involve stereochemical restrictions. In order to understand this better, we have been studying the effects of substituting other M^{2+} ions for the Zn^{2+} in our model systems whilst recognising that reproducing the stereochemical requirements of the protein may not always be possible.

Models for hydrolytic zinc enzymes

The interactions of carboxylates (e.g. from amino acid residues) and small substrate molecules with the metal site are important features of hydrolytic zinc enzymes. In order to model this the sterically demanding ligand potassium tris(3,5-dimethylpyrazolyl)hydroborate, $K^+[HB(C_5H_7N_2)_3]^-$ =KL, was used to provide a tri-coordinated $[LM]^+$ complex in which the pyrazolate moieties mimic imidazolate ligands.[3] This then reacts with anions such as NO_3^- (isoelectronic with CO_3^{2-}, and therefore of relevance to carbonic anhydrase[2]). It will also react with carboxylates such as acetate, which are known to be important for several zinc enzymes.[1] In previous work the even more sterically hindered ligands tris(3-tert-butyl-pyrazolyl)hydroborate (=L') and tris(3-phenylpyrazolyl)hydroborate (=L") have been used in this way.[4,5] The Cu(II) complex we have synthesised with L,[LCuOAc] **1**, can be isolated in high yields from ethanolic solutions containing $[Cu(OAc)_2.2H_2O]$ and KL. The structure of this compound determined by X-ray crystallography[3] is illustrated in Figure 1.

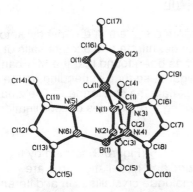

Figure 1   The molecular structure of **1**, [LCuOAc]

In **1** the acetate binds in a bidentate fashion with both Cu--O bond lengths almost equal within their standard deviations (1.929(5)Å and 2.040(5)Å). It is interesting to compare this behaviour with the previously reported [L'ZnOAc] where the acetate is monodentate.[4] Presumably it is either the greater steric bulk of this ligand which prevents the bidentate coordination of the carboxylate, or else an electronic effect in the case of the Cu(II) analogue which makes pentacoordination of the Cu(II) more important than the steric effect. The analogous [LCuNO$_3$] compound can be prepared by reacting Cu(NO$_3$)$_2$.3H$_2$O with KL in ethanol and once crystals suitable for X-ray diffraction have been obtained, it will be interesting to compare the bonding in this compound with that found for the related zinc compound [L"ZnNO$_3$].[5]

In early attempts to isolate **1** using CuCl$_2$ as a source of metal it was found that the presence of acetic acid accompanied by heating of the reaction mixture, resulted in the cleavage of the B--N bonds in the ligand and the formation of dimeric **2**, [CuCl$_2$dmpz$_2$]$_2$ (Hdmpz=3,5-dimethylpyrazole), as revealed by X-ray crystallography.[3]

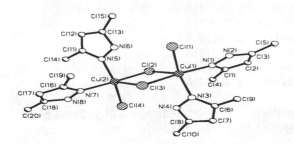

Figure 2   The molecular structure of **2**, [CuCl$_2$dmpz$_2$]$_2$

2 can also be prepared starting from CuCl$_2$ and dmpz in ethanol. The corresponding reaction starting from ZnCl$_2$ always results in the formation of the previously reported monomeric **3**, ZnCl$_2$dmpz$_2$.[6]

Activation of organic substrates by M(II)

We have been investigating the ligand system of di-2-pyridyl ketone $(py)_2CO$ in conjunction with the chloride, nitrate and acetate salts of divalent zinc, copper and nickel. It has been found that the M^{2+} can activate the carbonyl group on the ligand, so that, depending on the solvent used, water or ROH, diols and hemiketals may be formed. The zinc system proved to be the most versatile showing three modes of coordination and two different nuclearities.

If we consider the structures **4**, **5** and **6** (Figure 3) formed by the three metals in the NO_3^-/aq system, the zinc and nickel structures are isomorphous whereas the copper analogue crystallises in a different space group.[3] A comparison of the bond lengths reveals that the copper complex experiences Jahn-Teller distortion. In line with ligand field predictions, the nickel complex adopts regular octahedral coordination whereas the zinc is able to accommodate the restrictions of the ligand with distortions from octahedral symmetry.

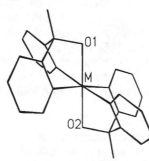
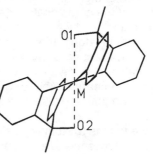

$M=Zn^{2+}$, Ni^{2+} $M=Cu^{2+}$

Bond lengths
Zn--O1	2.151(4)Å
Zn--O2	2.210(4)Å
Ni--O1	2.090(3)Å
Ni--O2	2.123(3)Å
Cu--O1	2.455(2)Å
Cu--O2	2.455(2)Å

Figure 3   A comparison of the coordination geometries of **4**, **5** and **6**
$[M((py)_2C(OH)_2)_2]^{2+}$ M=Zn, Ni, Cu

In contrast to this the structures of the copper and nickel complexes, **7** and **8** $[M((py)_2C(OH)(OHCH_3))_2]^{2+}$ in the OAc$^-$/aq system are isomorphous.[3] Again there is Jahn-Teller distortion in the copper complex. The nickel bond lengths are comparable to those in the nitrate system.

From the IR data we have reason to believe that in the $NiCl_2$ complexes formed in aqueous or alcoholic solution the ligand does not undergo any modification and is coordinated to the nickel through the pyridyl nitrogens as reported by Feller and Robson for the divalent chloride salts of Mn, Fe, Co, Ni and Cu.[7] The situation for zinc and copper is currently under investigation.

There appears to be an equilibrium process occurring in the zinc/EtOH or MeOH system between monomeric and dimeric forms. In the two dimers, **9** $[CH_3CO_2Zn(py)_2C(O)(OCH_3)]_2$ and **10** $[ClZn(py)_2C(O)(OC_2H_5)]_2$, isolated so far, the acetates in **9** adopt a cisoid configuration, as revealed by X-ray crystallography, whereas the chlorides in **10** were found to be transoid.[8]

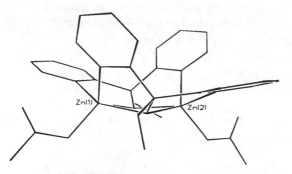

Figure 4   The structure of **9**, $[CH_3CO_2Zn(py)_2C(O)(OCH_3)]_2$, showing the cisoid arrangement of the acetate ligands.

Although the zinc is not crowded, the acetates are monodentate preserving pentacoordination in both dimers. In contrast to these and the hexacoordinate nitrate complex from the aqueous system, the zinc chloride monomer, **11** $[Cl_2Zn(py)_2C(OH)(OHC_2H_5)]$, is four coordinate with no coordination by the OH of the hemiketal group.[8] This shows the versatility of zinc to accommodate various modes of coordination and the ease with which it can vary coordination number. Both features are important properties for metals at catalytic sites.

Summary

1. The dipyridyl ketone/aqueous system yields products of general formula $[M\{(py)_2C(OH)_2\}_2]X_2$ for $X=NO_3^-$, OAc^-. Here the metals are hexa coordinate with Cu(II) showing significant Jahn-Teller distortion and Ni(II) having the geometry closest to octahedral. The situation for $X=Cl$ needs further investigation.
2. Zn(II) activates dipyridyl ketone in alcoholic media to produce coordinate hemiketal compounds in which Zn has variable coordination number and geometry.
3. The results with pyrazolyl borate based ligands suggest that whereas zinc seems to be adaptable in its modes of coordination, copper seems to

favour pentacoordination (e.g. **1**, **2**). This could be because it affords another means of relieving the degeneracy of unequally occupied d-orbitals.

References

1. R. J. P. Williams, Polyhedron, 1987, <u>6</u>, 61
2. R. Alsfasser, S. Trofimenko, A. Looney, G. Parkin, H. Vahrenkamp, Inorg. Chem., 1991, <u>30</u>, 4098
3. **1**: $CuC_{17}H_{25}N_6O_2B$, monoclinic P2(1)/c, a=13.983(2)Å, b=7.857(1)Å, c=18.943(3)Å, β=103.97(1)°; **2**: $Cu_2C_{20}H_{32}N_8Cl_4$, monoclinic P2(1), a=8.772(4)Å, b=13.802(5)Å, c=11.871(6)Å, β=106.39(4)°;
   4: $ZnC_{22}H_{22}N_6O_{11}$, monoclinic Cc, a=8.240(2)Å, b=31.845(5)Å, c=9.601(3)Å, β=96.57(2)°; **5**: $NiC_{22}H_{22}N_6O_{11}$, monoclinic Cc, a=8.183(1)Å, b=31.995(9)Å, c=9.532(2)Å, β=96.52(2)°;
   6: $CuC_{22}H_{24}N_6O_{12}$, monoclinic P2(1)/n, a=7.613(3)Å, b=11.978(5)Å, c=14.478(7), β=93.08(4)°; **7**: $CuC_{26}H_{30}N_4O_{10}$, monoclinic P2(1)/n, a=8.633(2)Å, b=7.903(2)Å, c=23.160(6)Å, β=98.03(2)°; **8**: $NiC_{26}H_{30}N_4O_{10}$ monoclinic P2(1)/n, a=8.845(3)Å, b=7.874(3)Å, c=22.843(5)Å, β=99.55(2)°; **9**: $ZnC_{28}H_{28}N_4O_8$, monoclinic P2(1)/n, a=8.320(2)Å, b=28.160(6)Å, c=13.351(3)Å, β=94.40(2)°.
4. R.Han, I. B. Gorrell, A. G. Looney, G. Parkin, J. Chem. Soc. Chem. Commun., 1991, <u>10</u>, 717
5. R. Alsfasser, A. K. Powell, H. Vahrenkamp, Angew. Chem. Int. Ed. Engl., 1990, <u>29</u>, 898
6. E. Bouwman, W. L. Driessen, R. A. G. De Graaff, J. Reedijk, Acta Cryst., 1984, <u>C40</u>, 1562
7. M. C. Feller, R. Robson, Aust. J. Chem., 1968, <u>21</u>, 2919
8. A. Abufarag, A. K. Powell, unpublished.

The Chemistry of Cadmium Chalcogenide Clusters [ECd$_8$(E'R)$_{16}$]$^{2-}$, [E$_4$Cd$_{10}$(E'R)$_{16}$]$^{4-}$, and [E$_4$Cd$_{17}$(E'R)$_{28}$]$^{2-}$

Ian Dance and Garry Lee

SCHOOL OF CHEMISTRY, UNIVERSITY OF NEW SOUTH WALES,
PO BOX 1, KENSINGTON, NSW 2033, AUSTRALIA

Cadmium sulfide and other metal chalcogenides (ME, E = S, Se, Te) are photoresponsive semiconductors. Visible light absorption excites an electron from a valence band to a conduction band, and the resulting high potential electron at ca -1.0v (NHE) and hole at ca +1.5v are able to perform redox chemistry: CdS and congeners are known to photocatalyse chemical processes and energy conversion, although not without problems. Further, the excitation energy (band gap) of CdS is dependent on the dimensions of the CdS crystallite within the size domain of 2-10 nm: the electronic structure of these nanocrystallites varies between the bonds of small molecules and the bands of the bulk crystal. There is therefore opportunity to tune the applications by size control.[1-4]

Surface structure is vital in mediating the photoredox chemistry of metal chalcogenides, and it is necessary to understand surface structure and dynamics in atomic detail. Nanocrystallites involve surface curvature with radii which are less than 10 times the Cd-S bond length, and this raises many questions about abnormal bonding at the surface, as shown diagrammatically in Figure 1.

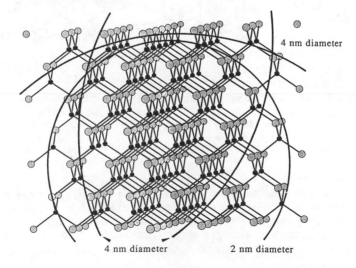

4 nm diameter

4 nm diameter   2 nm diameter

Figure 1. Part of the cubic lattice of CdS, with the surface curvature for 2 nm and 4 nm crystallites superimposed

As is often the case, molecular biology points to the answers for chemical problems. When grown with Cd^{2+} the yeast Candida glabrata generates CdS bionanocrystallites, and stabilises them against the thermodynamically favoured Ostwald growth.[5-7] We need to understand this chemistry.

Our approach to cadmium chalcogenide nanocrystallite chemistry has involved the growth of well-defined clusters, rather than the surfaces of colloids at the upper end of the size domain: we seek homogeneous and crystallisable compounds that can be characterised in atomic detail. The emphasis is on Cd, with incorporation of Se and Te, to take advantage of the favourable NMR characteristics of ^{113}Cd, ^{77}Se and ^{125}Te. We have concentrated on thiolate, selenolate and tellurolate (RE^-) as the surface ligands, because (i) they are manipulable substituted chalcogenides and thus the closest approximation to surface E^{2-}, (ii) they maintain the NMR advantage, and (iii) thiolate mimics the cysteine of the bionanocrystallites.

Three different $[E_wCd_x(E'R)_y]^z$ cluster structures are now confirmed by our research. All contain tetrahedrally coordinated Cd, tetrahedral or part-tetrahedral coordination at each E, E' atom, and overall tetrahedral topology and symmetry.

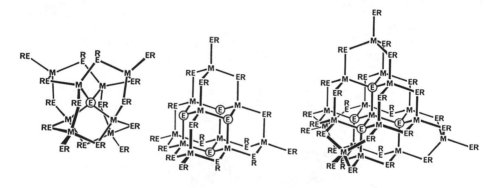

| I | II | III |

Structure **I**, $[ECd_8(E'R)_{16}]^{2-}$, contains a central E atom, coordinated to a tetrahedron of four inner Cd^i, each of which is bonded through three doubly-bridging E'R ligands to outer Cd^o, each of which has a terminal E'R ligand. The Cd^o atoms are arrayed as a tetrahedron, and the $(\mu$-E'R$)_{12}$ ligands constitute an icosahedron.[8]

Structure **II**, $[E_4Cd_{10}(E'R)_{16}]^{4-}$, is a fragment of the cubic (sphalerite, or zinc blende) lattice of bulk ME. There is an inner octahedron $(Cd^i)_6$ of which four triangular faces are capped with $(\mu_3$-E$)$ and the other four are capped with $(\mu$-E'R$)_3Cd^o$(E'R) groups. The $(\mu_3$-E$)_4$ array is a tetrahedron, the $(\mu$-E'R$)_{12}$ array is a truncated tetrahedron, and the $(Cd^o)_4$ array is a tetrahedron.[9]

Structure **III**, $[E_4Cd_{17}(E'R)_{28}]^{2-}$, contains an inner Cd^i bonded to four (μ_4-E), each of which is connected to three Cd^c such that the $(Cd^c)_{12}$ array is a cuboctahedron. Four of the triangular faces of this cuboctahedron are capped with $(\mu$-E'R$)_3Cd^o$(E'R) groups.[10]

Structures **I** and **III**, unlike **II**, are not fragments of non-molecular lattices. **II** contains four fused adamantanoid cages (characteristic of the cubic ME lattice), **I** contains no adamantanoid cages, **III** contains adamantanoid cages in the central region and barrelanoid cages (characteristic of the hexagonal ME lattice) at the caps. The structural principles characteristic of each structure type can be expanded, and it is possible to postulate an infinite series of clusters from each prototype. The compositions of the early members of each of these series are marked on the stoichiometry chart in Figure 2

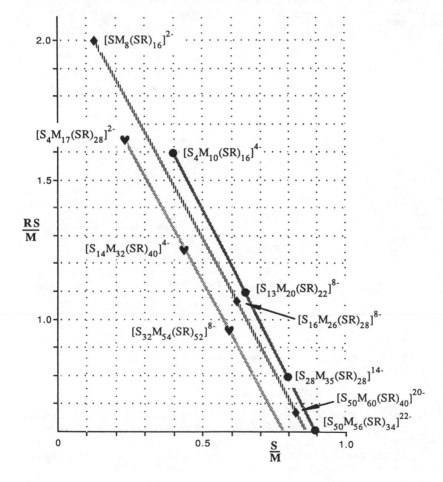

Figure 2. The compositions of the clusters in the three series

Synthesis

The challenge is the synthesis and crystallisation of these and similar clusters, with increasing sizes. The synthetic methods investigated are:

Self assembly of the components in homogeneous solution, reaction (1), is the fundamental method. It assumes thermodynamic control, and thermodynamic stability for the target molecules, selected according to the stoichiometry of the preparative solution. This method also requires knowledge of all factors such as the coordinative equilibria and the solvation influences, as well as solubility information (ie lattice packing energies).

(1) $Cd^{2+} + E^{2-}(HE^-) + E'R^- \rightarrow [E_wCd_x(E'R)_y]^z$

Kinetically controlled cluster formation occurs when the degree of association of Cd^{2+} and $E^{2-}(HE^-)$ is controlled physically, as for instance within micelles or zeolites.[3,11-13] This is called arrested precipitation.

Substitution of thiolate by chalcogenide, (2), can be attempted, as can substitution of chalcogenide by chalcogenol, equation (3)

(2) $[Cd_m(E'R)_n]^z + E^{2-}(HE^-) \rightarrow [E_wCd_x(E'R)_y]^z$

(3) $CdE + RE'H \rightarrow [E_wCd_x(E'R)_y]^z + H_2E$

Ligand and metal *exchange* reactions have been shown to generate many mixed ligand and mixed metal species in solution. Reactions of small clusters with metal electrophiles are expected to cause *condensation* to larger clusters: this possibility has been demonstrated in reactions such as (4).[14]

(4) $[S_4Cd_{10}(SPh)_{16}]^{4-} + Cd^{2+} \rightarrow [E_wCd_x(E'R)_y]^z$

Oxidised forms of the ligands can be used to advantage in redox syntheses. Thus the chalcogen E can be introduced as the element, or as the readily available soluble polychalcogenides $(E_p)^{2-}$, and these can be reduced with coordinated or uncoordinated $E'R^-$.

Accordingly, additional synthesis reactions are *self assembly with chalcogen reduction*, eq (5), reaction of metal thiolate with chalcogen, eq (6)

(5) $Cd^{2+} + E'R^- + E$ [or $(E_p)^{2-}$] $\rightarrow [E_wCd_x(E'R)_y]^z + RE'E'R$

(6) $[Cd_m(E'R)_n]^z + E$ [or $(E_p)^{2-}$] $\rightarrow [E_wCd_x(E'R)_y]^z + RE'E'R$

Thermolytic or photolytic *scission of the E–C bond* in chalcogenolates also leads to clusters $[E_wCd_x(E'R)_y]^z$

(7) $[Cd_m(E'R)_n]^z + h\nu \rightarrow [E_wCd_x(E'R)_y]^z$

Clusters with the four terminal ligands substituted with halide are prepared by the *oxidation* (8) as well as ligand exchange reactions

(8) $[E_wCd_x(E'R)_y]^z + 2X_2 \rightarrow [E_wCd_x(E'R)_{y-4}X_4]^z + 2RE'E'R$

Known Clusters

New clusters prepared by these methods and characterised by crystal structure determination and NMR are listed in Table 1

Table 1. Isolated and characterised clusters

Formula	E	E'	X	Xray	NMR
$[ECd_8(E'R)_{16}]^{2-}$	S	S		Me$_4$N$^+$	Cd
	S	Se		Me$_4$N$^+$ Et$_4$N$^+$	Cd, Se
	Se	Se			Cd, Se
	Te	Se			Cd
	S	Te			Cd
$[ECd_8(E'R)_{12}X_4]^{2-}$	S	S	Cl	Me$_4$N$^+$	
	S	Se	Br		Cd, Se
	S	Se	I		Cd, Se
$[E_4Cd_{10}(E'R)_{16}]^{4-}$	S	S		Me$_4$N$^+$ Et$_3$NH$^+$	Cd
	Se	S			Cd, Se
	Te	S		Me$_4$N$^+$	Cd
	S	Se			Cd, Se
	Se	Se			Cd, Se
$[E_4Cd_{10}(E'R)_{12}X_4]^{4-}$	S	S	Cl		Cd
	S	S	Br		Cd
	S	S	I	Me$_4$N$^+$	Cd
	Se	S	I		Cd
$[E_4Cd_{17}(E'R)_{28}]^{2-}$	S	S		Me$_4$N$^+$	Cd

R = phenyl and substituted phenyl

Self assembly equilibria, and new clusters

Investigations of self assembly equilibria define the relevant thermodynamic characteristics of homogeneous solutions, and bear upon the prospects for synthesis of the postulated larger clusters.   Solution compositions can be monitored by NMR.

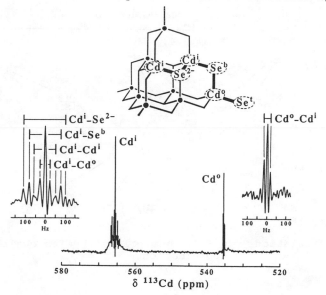

Figure 3 ^{113}Cd NMR specrum of $[Se_4Cd_{10}(SePh)_{16}]^{4-}$

Typical NMR spectra of a pure compound, and of the mixture of solids from a self assembly reaction, are shown in Figures 3 and 4. The cluster $[S_4Cd_{17}(SPh)_{28}]^{2-}$ undergoes reversible change in solution, but is NMR characterised in the solid state. We have defined the ^{113}Cd chemical shift correlations shown in Figure 5.

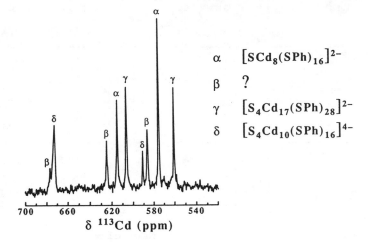

$\alpha \quad [SCd_8(SPh)_{16}]^{2-}$

$\beta \quad ?$

$\gamma \quad [S_4Cd_{17}(SPh)_{28}]^{2-}$

$\delta \quad [S_4Cd_{10}(SPh)_{16}]^{4-}$

Figure 4. Cd NMR of a Cd^{2+} / S^{2-} / PhS^- self assembly mixture

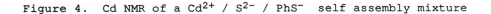
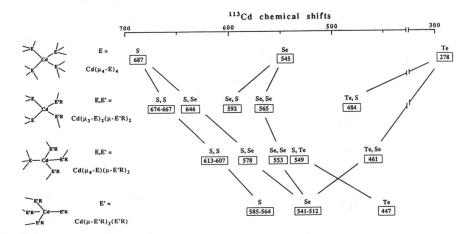

Figure 5. Correlations of ^{113}Cd chemical shift with chalcogenide coordination.

$$Cd^{2+} + PhS^- + S^{2-} \quad \text{in DMF}$$

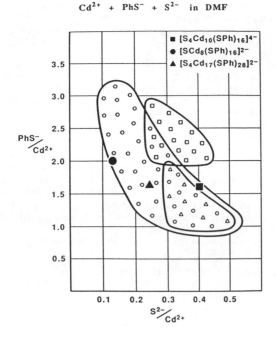

Figure 6.

Figure 6 shows the distribution of clusters present in self assembly solutions in the $Cd^{2+}/S^{2-}/SPh^-/DMF$ system. The conclusions from these investigations of homogeneous self assembly are that the three known cluster types **I, II** and **III** are the principal clusters and that additional species that form in solution must be small or be quite symmetrical with a small number of Cd sites per molecule.

Crystallisation of clusters from these solutions is not straightforward. For example, when self assembly reaction mixtures containing (according to NMR) only $[SCd_8(SPh)_{16}]^{2-}$ (or $[SCd_8(SPh)_{16}]^{2-}$ plus the $[S_4Cd_{17}(SPh)_{28}]^{2-}$ rearrangement product) are treated with $Me_4N^+Cl^-$ there is conversion to $[S_4Cd_{10}(SPh)_{16}]^{4-}$ only in solution, and it is this cluster which crystallises. The addition of a salt containing the isolation cation can modify the cluster equilibria. Further, it is necessary to remember that the anionic cluster which crystallises from solution is usually the least soluble anion/cation combination, not the most abundant cluster in solution. This factor is revealed also in the fact that solids precipitating from self assembly mixtures in acetonitrile can contain an unidentified cluster **IV** with [113]Cd resonances at 586, 624 and 676 ppm, while these lines are not observed in homogeneous solutions in DMF.

The general conclusion for the clusters $[SCd_8(SPh)_{16}]^{2-}$, $[S_4Cd_{10}(SPh)_{16}]^{4-}$ and $[S_4Cd_{17}(SPh)_{28}]^{2-}$ in the best known system is that the thermodynamic stabilities are in the order $[SCd_8(SPh)_{16}]^{2-} > [S_4Cd_{10}(SPh)_{16}]^{4-} > [S_4Cd_{17}(SPh)_{28}]^{2-}$, while the solubilities of the Me_4N^+ salts are $[S_4Cd_{10}(SPh)_{16}]^{4-} < [S_4Cd_{17}(SPh)_{28}]^{2-} < [SCd_8(SPh)_{16}]^{2-}$.[15]

Another difficulty encountered in the characterisation of these high-symmetry anionic clusters is the formation of crystals that are unstable to desolvation, twinned, or non-diffracting. Many cations, including cation "soups", are to be used in attempts at growth of suitable crystals. We are undertaking crystal engineering calculations of the lattice energies for crystals containing these clusters.

Conclusions: The protoype cluster types **I**, **II** and **III** have now been synthesised with variety, and characterised in solution and crystals. There are still two $[S_wCd_x(SPh)_y]^z$ clusters not yet characterised, namely **IV** and the species in equilibrium with crystalline $[S_4Cd_{17}(SPh)_{28}]^{2-}$. The prospects for systematic synthesis of the subsequent members of the postulated series are not positive for the methods investigated, and it is questionable whether they possess adequate thermodynamic advantage, in solution.

Acknowledgments This research is supported by the Australian Research Council. Garry Lee acknowledges an Australian Postgraduate Research Award.

REFERENCES

1. L.E. Brus *J. Phys. Chem.*, 1986, **90**, 2555-2560

2. A. Henglein *Chem. Rev.*, 1989, **89**, 1861-73

3. M.L. Steigerwald and L.E. Brus *Annu.Rev.Mater.Sci.*, 1989, **19**, 471-95

4. Y. Wang and N. Herron *J. Phys. Chem.*, 1991, **95**, 525-32

5. C.T. Dameron, R.N. Reese, R.K. Mehra, A.R. Kortan, P.J. Carroll, M.L. Steigerwald, L.E. Brus, and D.R. Winge *Nature*, 1989, **338**, 596-7.

6. C.T. Dameron, B.R. Smith, and D.R. Winge *J. Biol. Chem.*, 1989, **264**, 17355-60.

7. C.T. Dameron and D.R. Winge *Inorg. Chem.*, 1990, **29**, 1343-8.

8. G.S.H. Lee, K.J. Fisher, D.C. Craig, M.L. Scudder, and I.G. Dance *J. Am. Chem. Soc.*, 1990, **112**, 6435-7.

9. I.G. Dance, A. Choy, and M.L. Scudder *J. Am. Chem. Soc.*, 1984, **106**, 6285-6295.

10. G.S.H. Lee, D.C. Craig, I.N.L. Ma, M.L. Scudder, T.D. Bailey, and I.G. Dance *J. Am. Chem. Soc.*, 1988, **110**, 4863-4.

11. M.L. Steigerwald, A.P. Alivisatos, J.M. Gibson, T.D. Harris, R. Kortan, A.J. Muller, A.M. Thayer, T.M. Duncan, D.C. Douglass, and L.E. Brus *J. Am. Chem. Soc.*, 1988, **110**, 3046-3050.

12. N. Herron, Y. Wang, M.M. Eddy, G.D. Stucky, D.E. Cox, K. Moller, and T. Bein *J. Am. Chem. Soc.*, 1989, **111**, 530-540.

13. A.R. Kortan, R. Hull, R.L. Opila, M.G. Barwendi, M.L. Steigerwald, P.J. Carroll, and L.E. Brus *J. Am. Chem. Soc.*, 1990, **112**, 1327-32.

14. I.G. Dance and J.K. Saunders *Inorg. Chim. Acta*, 1985, **96**, L71-L73.

15. G.S.H. Lee, PhD thesis, University of NSW, 1992.

Structural Mis-matches in Silver and Gold Complexes of Thioether Macrocycles

A. J. Blake, R. O. Gould, C. Radek, G. Reid, A. Taylor, and M. Schröder*

DEPARTMENT OF CHEMISTRY, THE UNIVERSITY OF EDINBURGH, EDINBURGH EH9 3JJ, UK

1. INTRODUCTION

Macrocyclic ligands are well-known to form complexes that show remarkable thermodynamic stability and kinetic inertness.[1] The macrocyclic ligand can, therefore, be regarded as a protecting group for the metal centre controlling its stereochemical, electronic and redox properties. The introduction of inherent mis-matches between the metal and the donating ligands leads to the formation of strained complexes of unusual stereochemistry, with the resultant co-ordination geometries being a compromise between the preferences of the metal ion and the encapsulating ligand.[2] This has led to the stabilisation of mononuclear d^7 Ni(III),[3] Pd(III)[4] and Pt(III)[5] species such as $[M([9]aneS_3)_2]^{3+}$ and $[M([18]aneS_6)]^{3+}$. Related d^7 Rh(II)[4,6] and Ir(II)[7] have also been reported. We were interested in extending this work to structural and electrochemical studies of silver and gold complexes of thioether crowns.[8-10]

2. SILVER

Homoleptic Thioether Crowns

The structures of $[Ag([9]aneS_3)_2]^{+}$ [11] and $[Ag([18]aneS_6)]^{+}$ [8] have been reported previously and show homoleptic thioether co-ordination at octahedral Ag(I) centres. We have undertaken a study of the interaction of Ag(I) with $[15]aneS_5$ since this crown does not have sufficient S-donors to enable simple octahedral geometry at Ag(I). The structure of the $[Ag([15]aneS_5)]^{+}$ cation is indeed unusual and reflects the mismatch between Ag(I) and $[15]aneS_5$. The solid-state structure of the cation is dependent upon the counter-anion. Thus, the structure of $[Ag([15]aneS_5)]PF_6$ consists of parallel, polymeric chains of $[Ag([15]aneS_5)]^{+}$ units. The Ag(I) centre is bound to all five S-donors of the crown, Ag-S(1) = 3.219(5), Ag-S = 2.564(6), 2.659(5), 2.651(6), 3.075(7)Å (Figure 1). S(1) interacts with another Ag(I) centre, Ag'-S(1) = 2.742(5)Å, thus linking the $[Ag([15]aneS_5)]^{+}$ fragments to form the polymer chain. In contrast, the structure of $[Ag([15]aneS_5)]BPh_4$ shows (Figure 2) a dimeric structure in which two Ag(I) centres are bridged by two $[15]aneS_5$ ligands.[12] One Ag(I) is co-ordinated via one long and three short Ag-S bonds, while the other has one long and four short Ag-S contacts. However, [Ag([15]aneS$_5$)]B(C$_6$F$_5$)$_4$ shows (Figure 3) a genuine monomeric structure with five co-ordinate Ag(I), Ag-S = 2.4712(19), 2.5621(19), 2.7262(20), 2.6847(21), 2.8813(19)Å.

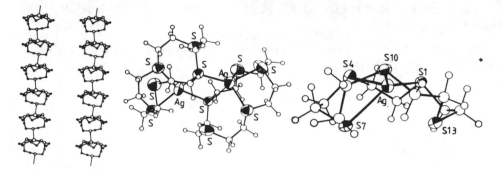

Figure 1   Figure 2   Figure 3

Mixed Oxy-Thioether Crowns

The extraction and transport of Ag(I) and related heavy metal ions across liquid-liquid interfaces and membranes have been achieved using polyether and polythioether cyclic ionophores.[13] The complexation and transport of Ag(I) has focussed particularly on the use of mixed S- and O-donor crowns. The precise mode of co-ordination of these mixed-donor macrocycles to Ag(I) is unknown although 1:1 and 2:1 ligand:Ag complexes involving *exo*-binding of S-donors have been postulated. We were interested in determining the precise mode of co-ordination of the simple mixed O/S macrocycles [15]aneS$_2$O$_3$ [14,15] and [18]aneS$_2$O$_4$ (L) to Ag(I).

[15]aneS$_2$O$_3$ [18]aneS$_2$O$_4$

Reaction of Ag(I) with one molar equivalent of L affords the expected 1:1 complexes [Ag(L)]$^+$. However, the solid state structure of [Ag([15]aneS$_2$O$_3$)]$^+$ shows (Figure 4) a polymeric chain structure with [Ag([15]aneS$_2$O$_3$)]$^+$ fragments connected by bridging thioether S-donors. Reaction of Ag(I) salts with [15]aneS$_2$O$_3$ in ratios of 1:1.5 to 1:3 affords the binuclear species [Ag$_2$([15]aneS$_2$O$_3$)$_3$]$^{2+}$ (Figure 5). Interestingly, the solid-state structure of [Ag([18]aneS$_2$O$_4$)]$^+$ is more complicated than expected and shows a binuclear structure incorporating bridging S-donors. Each [Ag$_2$([18]aneS$_2$O$_4$)$_2$]$^{2+}$ unit is linked to another *via* a long-range S····S interaction

(3.434Å) to give an overall tetrameric structure of stoichiometry $[Ag_4([18]aneS_2O_4)_4]^{4+}$ (Figure 6).

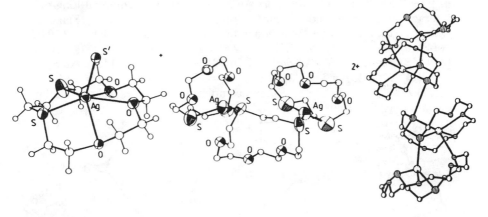

Figure 4   Figure 5   Figure 6

3. GOLD

Au(I) is well known to prefer linear co-ordination; however, treatment of $[Au(tht)_2]^+$ (tht = tetrahydrothiophene) with [9]aneS$_3$, [15]aneS$_5$, [18]aneS$_6$ and Me$_2$[18]aneN$_2$S$_4$ leads to the formation of complexes exhibiting highly unusual geometries. The single crystal X-ray structure of $[Au([9]aneS_3)_2]^+$ shows (Figure 7) one [9]aneS$_3$ co-ordinated as a monodentate ligand and is therefore relatively labile, Au-S(1') = 2.302(6)Å, while the other [9]aneS$_3$ is bound asymmetrically with one short and two long Au-S distances, Au-S(1) = 2.350(7), Au-S(4) = 2.733(8), Au-S(7) = 2.825(8)Å, <S(1')-Au-S(1) = 153.98(23)°.[10] Thus, the Au(I) centre in $[Au([9]aneS_3)_2]^+$ shows [2+2] S-co-ordination and can therefore be regarded as being in a tetrahedrally-distorted linear environment.[10] The structures of the corresponding Au(II) and Au(III) complexes $[Au([9]aneS_3)_2]^{2+}$ (Figure 8) and $[Au([9]aneS_3)_2]^{3+}$ (Figure 9) reflect the compromise between the facially pre-organised [9]aneS$_3$ crown and the preferred stereochemistries of d^9 and d^8 metal ions.

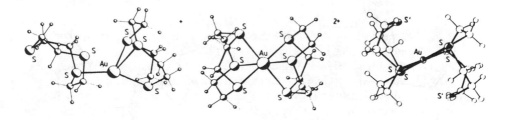

Figure 7   Figure 8   Figure 9

The related complexes $[Au([18]aneS_6)]^+$ (Figure 10) and $[Au(Me_2[18]aneN_2S_4)]^+$ (Figure 11) show similar geometries at Au(I) to those in $[Au([9]aneS_3)_2]^+$ with two short Au-S distances and additional long-range interactions to other donors within the ring. The structure of $[Au([18]aneS_6)]^+$ shows two conformers for the cation in the solid state with Au-S(1) = 2.321(3), Au-S(10) = 2.320(4), Au-S(7) = 2.856(4), Au-S(4) = 2.870(4)Å; for $[Au(Me_2[18]aneN_2S_4)]^+$, Au-S(1) = 2.304(2), Au-S(10) = 2.314(2), Au-S(4) = 2.888(2), Au···S(13) = 3.624(2), Au···N(7) = 3.795(5), Au···N(16) = 3.660(5)Å, <S(1)-Au-S(10) = 177.30(5)°. These complexes, therefore, incorporate large chelate rings that span from 150° to almost 180°.

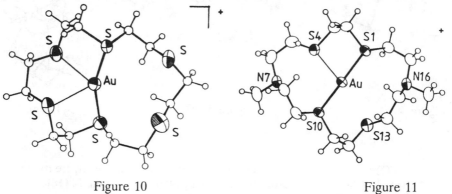

Figure 10   Figure 11

As for the Ag(I) complexes of $[15]aneS_5$, the Au(I) species $[Au([15]aneS_5)]^+$ shows unusual structural features. Encapsulation of Au(I) by $[15]aneS_5$ to give the same structure as that observed in $[Au([18]aneS_6)]^+$ would be unlikely given the inherent strain that would be imposed on the 15-membered ring crown to form a 155° chelate as in Figure 10. The single crystal X-ray structures of $[Au([15]aneS_5)]PF_6$ and $[Au([15]aneS_5)]B(C_6F_5)_4$ both show the cations to be dimeric with Au(I) centres bridging $[15]aneS_5$ crowns in a macrobicyclic structure (Figure 12); in $[Au([15]aneS_5)]PF_6$, for example, Au-S(1) = 2.345(7), Au-S(10') = 2.223(5), Au-S(4) = 2.887(6), Au-S(13) = 2.992(6)Å, S(1)-Au-S(10') = 158.82(22)°.

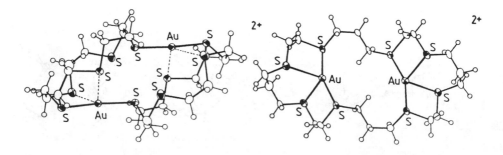

Figure 12   Figure 13

Reaction of two molar equivalents of $[Au(tht)_2]^+$ (tht = tetrahydrothiophene) with one molar equivalent of [24]aneS$_8$ or [28]aneS$_8$ (L) affords the corresponding binuclear complexes $[Au_2(L)]^{2+}$. The analogous binuclear Ag(I) complexes can also be prepared. The single crystal X-ray structure of $[Au_2([28]aneS_8)]^{2+}$ shows (Figure 13) two Au(I) centres complexed tetrahedrally within the octathia macrocycle in a [2+2] manner, Au-S(1) = 2.3301(19), Au-S(11) = 2.3378(18), Au-S(4) = 2.7891(20), Au-S(8) = 2.7629(20), Au···Au = 5.6977(6)Å. The <S(1)-Au-S(11) angle of 155.58(6)° contrasts with the value of 133.39(20)° observed previously for $[Cu_2([28]aneS_8)]^{2+}$ [16] and reflects the preferences of Au(I) for linear and Cu(I) for tetrahedral co-ordination.

The oxidation of $[Au([9]aneS_3)_2]^+$, $[Au([15]aneS_5)]^+$, $[Au([18]aneS_6)]^+$ and $[Au(Me_2[18]aneN_2S_4)]^+$ to form mononuclear Au(II) and Au(III) species has been monitored spectroelectrochemically using *in situ* electronic and esr spectroscopy. Table 1 summarises cyclic voltammetric and coulometric data for these Au(I) thioether complexes.

TABLE 1: **Cyclic Voltammetric[a] and Coulometric Data for Au(I) Thioether Complexes**

Complex	$^1E_{1/2}$	n[b]	$^2E_{1/2}$	n[b]
$[Au([9]aneS_3)_2]^+$	+0.12[i,c]	1	+0.46	1
$[Au([15]aneS_5)]^+$	+0.36	1	+0.54	1
$[Au([18]aneS_6)]^+$	+0.36	1	+0.56	1
$[Au(Me_2[18]aneN_2S_4)]^+$	+0.14	1	+0.43	1
$[Au([18]aneN_2S_4)]^+$	+0.85[i]	2		
$[Au_2([28]aneS_8)]^{2+}$	+0.55[br]	4		
$[Au([16]aneS_4)]^+$	+0.14[br]	2		

i: irreversible; a: measured in MeCN (0.1M $^nBu_4NPF_6$) at 293K at platinum electrodes at a scan rate of 100mV.sec^{-1}. Potentials quoted in V vs Fc/Fc$^+$; b: number of electrons per cation obtained coulometrically at a Pt basket at 243K; br: broad wave with peak-to-peak separation of 200mV.sec^{-1}; c: irreversible wave (E_{pc}) due to slow Au(I) → Au(II) electron-transfer.

The one-electron oxidation products are esr-active and show four-line spectra in solution due to hyperfine coupling to ^{197}Au (I = 3/2, 100%). Interestingly, the complexes of homoleptic thioether crowns that incorporate non-interacting S-donors in the precursor Au(I) species afford the most stable Au(II) species. This reflects the requirement of increasing co-ordination number on going from Au(I) (two co-ordinate) to Au(II) (six co-ordinate, see below). Thus, these precursor Au(I) complexes may be regarded as stereochemically strained in the sense that the extra S-donors destabilise the d^{10} Au(I) configuration and provide a driving force for the formation of mononuclear Au(II). Esr spectroscopic data for the Au(II) complexes in fluid solution are summarised in Table 2. The esr spectra of the Au(II) species in frozen glass are much more complicated and current work is aimed at simulating and interpreting these spectra to obtain accurate g-values.

TABLE 2: *X*-Band Esr Spectral Data (293K, MeNO$_2$ solution) for Mononuclear Au(II) Thioether Complexes

Complex	g_{iso}	A_{iso}
[Au([9]aneS$_3$)$_2$]$^{2+}$	2.016	44.3G
[Au([15]aneS$_5$)]$^{2+}$	2.014	43.3G
[Au([18]aneS$_6$)]$^{2+}$	2.026	45.5G

ACKNOWLEDGEMENTS

We thank SERC for support, and the Royal Society of Edinburgh and Scottish Office Education Department for a Support Research Fellowship to MS.

REFERENCES

1.   *The Chemistry of Macrocyclic Ligand Complexes*, L.F. Lindoy, Cambridge University Press, Cambridge, UK, 1989.
2.   A.J. Blake and M. Schröder, *Advances in Inorganic Chemistry*, 1990, **35**, 1; G. Reid and M. Schröder, *Chemical Society Reviews*, 1990, **19**, 239; M. Schröder, *Pure and Applied Chemistry*, 1990, **60**, 517.
3.   A.J. Blake, R.O. Gould , M.A. Halcrow, A.J. Holder and M. Schröder, *J. Chem. Soc., Dalton Transactions*, in press; A.J. Holder, PhD Thesis, University of Edinburgh, 1987.
4.   A.J. Blake, A.J. Holder, T.I. Hyde and M. Schröder, *J. Chem. Soc., Chem. Commun.*, 1987, 987.
5.   A.J. Blake, R.O. Gould, A.J. Holder, T.I. Hyde, M.O. Odulate, A.J. Lavery and M. Schröder, *J. Chem. Soc., Chem. Commun.*, 1987, 118.
6.   S.C. Rawle, R. Yagbasan, K. Prout and S.R. Cooper, *J. Am. Chem. Soc.*, 1987, **109**, 6181; A.J. Blake, R.O. Gould, A.J. Holder, T.I. Hyde and M. Schröder, *J. Chem. Soc., Dalton Trans.*, 1988, 1861; S.R. Cooper, S.C. Rawle, R. Yagbasan and D.J. Watkin, *J. Am. Chem. Soc.*, 1991, **113**, 1600.
7.   A.J. Blake, R.O. Gould, A.J. Holder, T.I. Hyde and M. Schröder, *J. Chem. Soc., Dalton Trans.*, 1990, 1759.
8.   A.J. Blake, R.O. Gould, A.J. Holder, T.I. Hyde and M. Schröder, *Polyhedron*, 1989, **8**, 513 and references therein.
9.   A.J. Blake, G. Reid and M. Schröder, *J. Chem. Soc., Dalton Trans.*, 1991, 615.
10.   A.J. Blake, R.O. Gould, J.A. Greig, A.J. Holder, T.I. Hyde and M. Schröder, *J. Chem. Soc., Chem. Commun.*, 1989, 876; A.J. Blake, J.A. Greig, A.J. Holder, T.I. Hyde, A. Taylor and M. Schröder, *Angew. Chem.*, 1990, **102**, 203; *Angew. Chem., Int. Ed. Engl.* 1990, **29**, 197 and references therein.
11.   H-J. Küppers, K. Wieghardt, Y-H. Tsay, C. Krüger, B. Nuber and J. Weiss, *Angew. Chem.*, 1987, **99**, 583; *Angew. Chem., Int. Ed. Engl.*, 1987, **27**, 575; J. Clarkson, R. Yagbasan, P.J. Blower, S.C. Rawle and S.R. Cooper, *J. Chem. Soc., Chem. Commun.*, 1987, 950.
12.   A.J. Blake, R.O Gould, G. Reid and M. Schröder, *J. Chem. Soc., Chem. Commun.*, 1990, 974.

13.   R.M. Izatt, L. Eblerhardt, G.A. Clark, R.L. Bruening, J.S. Bradshaw and J.J. Christensen, *Separation Science and Technology*, 1987, **22**, 701 and references therein; E. Sekido, K. Saito, Y. Naganuma and H. Kumazaki, *Analytical Sciences* 1985, **1**, 363; M. Oue, K. Kimura and T. Shono, *Analytica Chimica Acta*, 1987, **194**, 293; K. Chamaya and E. Sekido, *Analytical Sciences*, 1987, **3**, 535; D. Lamb, R.M. Izatt, J.J. Christensen and D.J. Eatough, in *Coordination Chemistry of Macrocyclic Compounds*, Ed G.A. Melson, Plenum Press, 1979, p173 and p158.

14.   A.J. Blake, G. Reid and M. Schröder, *J. Chem. Soc., Dalton Transactions*, 1990, 3849.

15.   A.J. Blake, G. Reid and M. Schröder, *J. Chem. Soc., Chem. Commun.*, 1992, in press.

16.   A.J. Blake, A. Taylor and M. Schröder, *Polyhedron*, 1990, **9**, 2911.

Group 12 Complexes of Tetrakis(2-aminoethyl)cyclam

Ling H. Tan[1], Max R. Taylor[1], Kevin P. Wainwright[1,*], and Paul A. Duckworth[2]

[1] SCHOOL OF PHYSICAL SCIENCES, THE FLINDERS UNIVERSITY OF SOUTH AUSTRALIA, GPO BOX 2100, ADELAIDE, SOUTH AUSTRALIA 5001, AUSTRALIA

[2] DEPARTMENT OF PHYSICAL AND INORGANIC CHEMISTRY, UNIVERSITY OF ADELAIDE, GPO BOX 498, ADELAIDE, SOUTH AUSTRALIA 5001, AUSTRALIA

1 INTRODUCTION

Although extensive work has been carried out on the transition metal complexes of tetrakis(2-aminoethyl)cyclam (taec),[1] no attention has previously been paid to the possibility of forming diamagnetic complexes from Zn(II) and Cd(II).

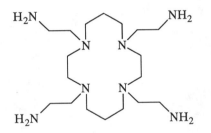

(taec)

It was our original intention to form these complexes and use them in variable temperature ^{13}C NMR studies directed towards an investigation of the conformational changes that are thought to occur.[1] However, as it turned out, the low solubility of the complexes precluded this work, but during the course of the synthesis and subsequent physical measurements we noted significant differences between these d^{10} systems and the transition metal complexes.

2 RESULTS AND DISCUSSION

Complexation with Zinc(II)

Potentiometric titration of fully protonated taec zinc(II) mixtures with base gives rise to the speciation shown in Figure 1. It is evident from the Figure that, irrespective of the zinc : ligand ratio utilised, the chemistry of the system is dominated by the tendency of the ligand to bind two zinc ions and for the complex to undergo hydrolysis such that a single hydroxyl group bridges the two

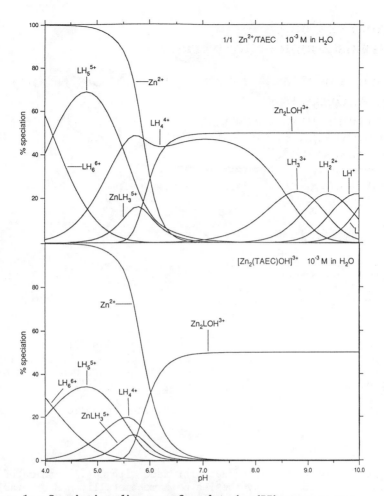

Figure 1   Speciation diagrams for the zinc(II) taec system

metal ions. Our measurements give a value of 5.8 as the upper limit for the pK$_a$ of the, presumed, co-ordinated water molecule which acts as precursor to the bridge. The formation of a hydroxyl bridge in aqueous conditions has been noted previously only with the di-cobalt(II) complex,[1] but whereas in that case the hydroxyl bridge may be substituted by other potential bridging anions such as Cl$^-$ and Br$^-$ this will not occur with zinc(II).

[Zn$_2$(OH)(taec)][ClO$_4$]$_3$ was prepared using standard techniques and through a variety of experiments it was verified that this was the only complex cation that could be isolated. X-ray crystallography indicated that its structure (shown in Figure 2) is similar to that of the di-cobalt(II) species: Each zinc ion has approximately square pyramidal co-ordination geometry and is displaced slightly (0.671(4) & 0.710(4) Å for Zn(1) and Zn(2) respectively) from the mean basal plane formed by two nitrogen atoms from the macrocyclic ring and two pendant amino groups, in the direction of the bridging hydroxyl group. The two zinc ions are separated by 3.73(1) Å and the two Zn-O bonds are

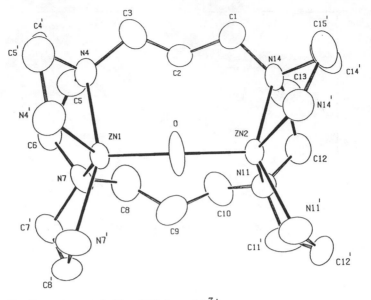

Figure 2   Structure of $[Zn_2(OH)(taec)]^{3+}$

the same, being measured at 1.938(5) and 1.931(5) Å for Zn(1)-O and
Zn(2)-O, respectively. The Zn(1)-O-Zn(2) angle is 149.8(3)°. The
bridging hydroxide ion forms intramolecular hydrogen bonds with all
four pendant amino groups [O-N(4') = 3.195(9), O-N(7') = 3.181(9),
O-N(11') = 3.149(9) and O-N(14') = 3.295(9) Å] and intermolecular
hydrogen bonds with a neighbouring perchlorate anion [O-O = 3.11(1)
and O-O' = 3.14(1) Å].

Complexation with Cadmium(II)

Potentiometric titration data for taec with Cd(II) indicated that
it should be possible to synthesise both monometallic and bimetallic
species (Figure 3 shows speciation diagrams for 1 : 1 and 2 : 1, metal
ion : taec, mixtures). This was verified through synthesis and
isolation of $[Cd(taec)][ClO_4]_2$ and $[Cd_2(taec)][ClO_4]_4$. The former
species is, to date, unique in so far as it is the only substantiated
monometallic species observed with this ligand. It behaves as a 2 : 1
electrolyte in DMF and shows a distinctly different ^{13}C NMR spectrum
to the bimetallic species. The latter species will add a bridging
anion readily (the upper limit for the pK_a of the, presumed,
co-ordinated water molecule that leads to a bridging hydroxyl is 6.8)
and it proved possible to isolate and grow crystals, satisfactory
for X-ray analysis, of $[Cd_2(Cl)(taec)][CF_3SO_3]_3$. The structure (shown
in Figure 4) is similar to that with various transition metal ions:
Both cadmium atoms exhibit distorted square pyramidal stereochemistry
with the bridging halide ion at the apical site and a basal plane
consisting of a pair of 1,4 related ring nitrogen atoms and two
pendant amino groups. The Cd(1)-Cl and Cd(2)-Cl distances are similar
being 2.521(3) and 2.544(3) Å respectively. The inter-cadmium
separation is 4.41(2) Å and the two ions are both displaced from their
basal plane towards the bridging halide by 0.803(6) and 0.834(6) Å
respectively. The Cd(1)-Cl-Cd(2) angle is 121.0(1)°.

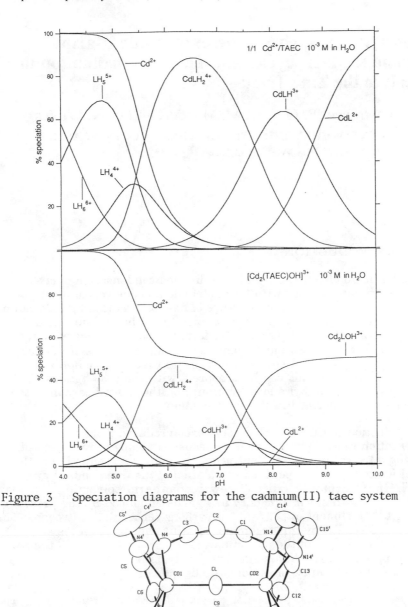

Figure 3 Speciation diagrams for the cadmium(II) taec system

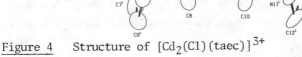

Figure 4 Structure of $[Cd_2(Cl)(taec)]^{3+}$

REFERENCE

1. See, for example, I. Murase, I. Ueda, N. Marubayashi, S. Kida, N. Matsumoto, M. Kudo, M. Toyohara, K. Hiate and M. Mikuriya, *J.Chem.Soc.,Dalton Trans.*, 1990, 2763 and references cited therein.

Copper(II)-directed Syntheses of Pendant-arm Polyamine Macrocycles and Their Co-ordination to Metals in the Zinc Triad

G. A. Lawrance, P. G. Lye, M. Maeder, and E. N. Wilkes

DEPARTMENT OF CHEMISTRY, THE UNIVERSITY OF NEWCASTLE, CALLAGHAN, NSW 2308, AUSTRALIA

1 INTRODUCTION

Copper(II) is an effective template for directed condensation of *cis*-disposed primary amines with formaldehyde and the carbon acid nitroethane in basic solution. A range of new polydentate ligands can be prepared in a single step, the reaction facility not being influenced greatly by variation in donors remote from the pair of primary amines involved in the condensation.[1,2] The reactions proceed essentially as described in Figure 1. Using an appropriate precursor complex, macromonocyclic ligands can be prepared, with the new ligands carrying pendant methyl and nitro groups on the central carbon of the new chelate ring formed in the condensation.

Reduction of the copper(II) complexes produced from the condensation reaction with zinc in aqueous acid leads to reduction of both the nitro group to a pendant primary amine, and reduction of the copper ion to the metal, permitting isolation of the free pendant-arm ligand. In particular, a range of new pendant-arm polyamine macrocycles, with steric efficiency as ligands enhanced by the procedure leading to attachment of pendants at carbons rather than nitrogens, can be prepared readily by this technique. In this paper, syntheses and complexation studies of the polyamines (1) to (6) with metals in the zinc triad in particular are discussed. These polyamines have the potential to act as quinquedentate or sexidentate ligands.

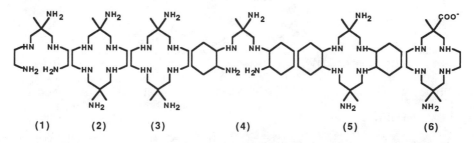

(1)   (2)   (3)   (4)   (5)   (6)

2 EXPERIMENTAL METHODS

Syntheses of new complexes and ligands employed procedures essentially like those previously described for (1),(2) and (3),[3,4] except reactions using *trans*-cyclohexane-1,2-diamine (chxn) were performed in aqueous solution rather than in methanol due to limited solubility of reactant complexes in the latter solvent. For reactions with chxn, resolution of the ligand with (+)-tartrate permitted reaction with *RR, SS* or racemic chxn, with both (4) and (5) products isolable. Further reaction of the acycle (1) with formaldehyde and the different carbon acid diethylmalonate permits eventual isolation of the polyaminoacid (6). Spectroscopic methods employed have been described previously,[3,4] and potentiometric titration facilities and methods employed were those reported earlier.[5]

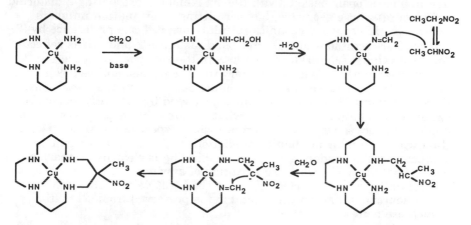

Figure 1. Mechanism for the copper(II)-directed condensation reaction.

3 RESULTS AND DISCUSSION

Copper(II)-directed condensation reactions employing ethane-1,2-diamine (en) produce both the acyclic analogue (1) where condensation has occurred at only one pair of *cis*-disposed primary amines, and the macromonocycle (2). When the precursor diamine is varied by successive methyl substitution on one of the carbons of en, the ratio of acycle:macrocycle rises from *ca.* 1:4 for en to 3:1 for propane-1,2-diamine to 25:1 for 2-methyl-propane-1,2-diamine. This suggests a subtle steric effect on macrocyclization; isolated acycles products appear to be the isomer with the methyl substituents exclusively adjacent to the primary amines. With chxn, both acyclic (4) and cyclic (5) products are also obtained. The ratio of acycle:macrocycle differs depends on whether *RR,RR (SS,SS)* or *RR,SS* bis(diamine) precursors are employed, the latter yielding dominantly the macrocycle. This may relate to differing steric demands of the *RR,RR* or *RR,SS* reactants and products, tied to differering chelate ring conformations of the chxn units in each case. Molecular mechanics analyses of the various isomers when coordinated to metal ions identifies different strain for different optical isomers. The

[13]C n.m.r. spectra of the free ligands reflect the lower symmetry of the *RR,RR* species, showing splitting of single resonances in the *RR,SS* analogue as a result of non-equivalent magnetic environments for the two halves of the molecule.

Formation of the macrocycle from reaction of en occurs to produce two geometric isomers, with the two pendant primary amines either *anti* (2) or *syn* (3), the former the major isomer.[4] The stereospecificity in the condensation has been related to weak axial interaction of nitro groups during condensation, since a partially condensed intermediate with two nitro groups interacting axially can yield only the *anti* isomer. The co-ordination mode of metal ions to the *anti* and *syn* isomers differs significantly (Figure 2)[4,6], the former effectively encapsulating the metal ion in a distorted octahedral environment, the latter binding the metal ion in a more open 'basket', with the six N-donors presenting a distorted trigonal prismatic geometry, but permitting co-ordination numbers above six. Molecular mechanics calculations predict a preference by (2) for smaller metal ions and shorter bond distances,[7] and this is reflected in log K values (Table 1) for metal ions in the zinc triad (and the additional metal ion lead) which vary in ionic radius, but, as d^{10} systems, make no formal stereochemical demands on the metal ions. The small zinc(II) ion is significantly preferred by the *anti* isomer (2), whereas, the larger ions are somewhat favoured by the *syn* isomer (3), which in effect shows little preference with respect to metal ion size, in line with molecular mechanics predictions.[7] That there is a general influence of the pendant groups on the binding is clear by comparing the results of (2) and (6), where substitution of a primary amine group by a carboxylate group has a clear effect on the log K values, although the same approximate trend with metal ion is preserved from (2) to (6), which are both anti isomers.

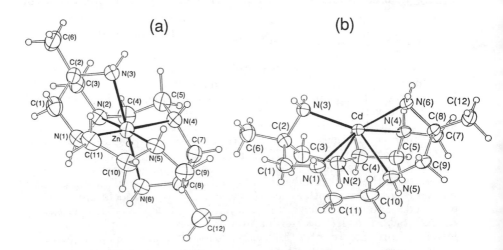

Figure 2. Structural examples of coordination of (a) *anti* (2) and (b) *syn* (3) isomers to metal ions in the zinc triad.[4,6]

Table 1.   Log K values for $M^{2+} + L = ML^{2+}$ with (2), (3) and (6).

Metal Ion	Free-Ion Radius, Å	(2)	(3)	(6)
Zn^{2+}	0.74	15.0	10.4	14.7
Cd^{2+}	0.97	10.6	12.0	10.9
Hg^{2+}	1.10	10.5	12.2	7.8
Pb^{2+}	1.20	10.8	11.1	8.6

The marked fall in stability for the *anti* isomer (2) with increasing metal ion size presumably relates to difficulty in this pseudo-encapsulating ligand accommodating the larger metal ions, with their preferred longer bonds [e.g. Zn-N typically 2.20 Å compared with Cd-N 2.38 Å and Hg-N 2.5 Å] in the relatively small cavity. The step in log K from zinc(II) to cadmium(II) is consistent with our inability to isolate a sexidentate complex of cadmium(II) or larger ions, although these form readily with the small zinc(II) and other small metal ions. This view is reinforced by our recent isolation and structural characterization of mercury(II) complexes of (2) in which the mercury binds outside the macrocyclic cavity, either with (2) co-ordinated only via a single bond from a primary pendant amine, or with chelation of (2) involving a primary pendant amine and one of the secondary amines. The co-ordination of metal ions involving the pendant amines but not the full macrocycle donor set may also present an understanding of the kinetics of coordination of copper(II) to (2), where two sequential steps are observed, rather than the single process reported for the unsubstituted tetraazacycloalkane parent cyclam. The fast first step may involve formation of an intermediate with the copper ion bound as an 'external' chelate, with the copper then 'stepping' into the macrocycle ring in the second slower process. The pendant in this way is facilitating binding, even though in the case of copper(II) the thermodynamically stable product does not involve binding by the pendant primary amines.

REFERENCES

1. P. V. Bernhardt and G. A. Lawrance, Coord. Chem. Rev., 1990, 93, 297.
2. P. Comba, T. W. Hambley, G. A. Lawrance, L. L. Martin, P. Renold and K. Varnagy, J. Chem. Soc., Dalton Trans., 1991, 277.
3. P. Comba, N. F. Curtis, G. A. Lawrance, A. M. Sargeson, B. W. Skelton and A. H. White, Inorg. Chem., 1986, 25, 4260.
4. P. V. Bernhardt, P. Comba, T. W. Hambley, G. A. Lawrance and K. Varnagy, J. Chem. Soc., Dalton Trans., 1992, 355.
5. G. K. Hollingshed, G. A. Lawrance, M. Maeder and M. Rossignoli, Polyhedron, 1991, 10, 409.
6. P. V. Bernhardt, G. A. Lawrance, M. Maeder, M. Rossignoli and T. W. Hambley, J. Chem. Soc., Dalton Trans., 1991, 1167.
7. P. V. Bernhardt and P. Comba, Helv. Chim. Acta, 1991, 74, 1834.

Stereoselective Nitrosations of Pendant Amine Substituted Macrocyclic Copper(II) Complexes

Paul V. Bernhardt, Peter Comba, and Catherine M. Mitchell

INSTITUT FÜR ANORGANISCHE CHEMIE, UNIVERSITÄT BASEL,
SPITALSTRASSE 51, 4056 BASEL, SWITZERLAND

Introduction

Nitrosation of aliphatic primary amines with sodium nitrite in dilute aqueous acid is generally a rather intractable and hence seldom recommended synthetic procedure.[1] Nevertheless, we have found that performing such reactions on pendant primary amino-substituted macrocyclic complexes may lead to a remarkable stereoselectivity, in particular the formation of exocyclic instead of endocyclic C=C bonds as elimination products.

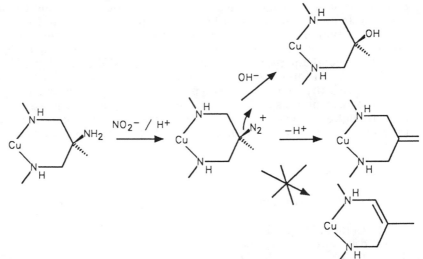

Results and Discussion

If the reaction is conducted in aqueous solution, and at low concentrations of competing nucleophiles, the products of the reaction are as shown above. In all cases studied so far, the formation of the exocyclic (as opposed to endocyclic) olefinic groups has been exclusive, however the ratio of elimination (alkenes) to substitution (alcohols) products varies from system to system, and in the case of 6 no elimination product was identified.

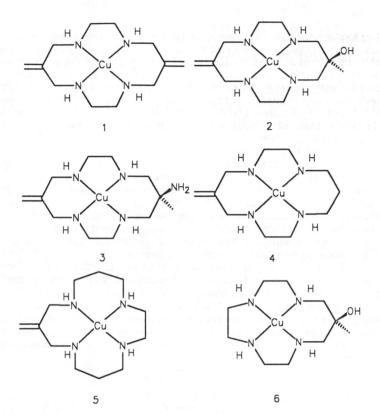

The assignment of exocyclic olefinic groups has been made based on NMR spectra of the metal-free ligands, achieved by precipitation of Cu(II) with sulfide. In all cases investigated so far, the exocyclic C=C bond has been identified by two well separated ^{13}C resonances at *ca*. 120 and 140 ppm corresponding to the respective terminal and fully substituted olefinic C atoms, and an IR resonance for the C=C bond at *ca*. 1655 cm^{-1}. The same nitrosation reaction may be performed on the analogous Pd(II) complex of **1**, which allowed an NMR analysis of the reaction solution, without disturbing the mixture by precipitation of the metal ion. This spectrum revealed not only an exclusive formation of exocyclic C=C bonds, but also a high selectivity towards elimination versus substitution, i.e. the major product was the Pd(II) analogue of **1** with a relatively small amount of the hydrated complex **2** being identified (Figure).

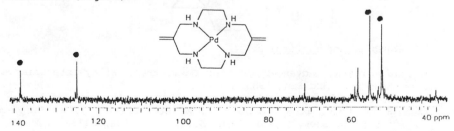

Figure. ^{13}C NMR Spectrum of Reaction Mixture

The stereoselectivity in the present reactions is quite remarkable when one surveys the generally diverse range of products that may emerge from nitrosations of aliphatic primary amines.[2] In addition to elimination and substitution reactions, rearrangements are commonplace; however such processes are yet to be observed with the present systems. The chelate ring evidently acts to direct the stereochemistry toward formation of an exocyclic double bond (which is inherently less stable than the more substituted Saytzeff product), and also appears to hold the (presumed) carbocation intermediate in place so that no rearrangement takes place. However, similar reactions performed on pendent amino substituted macrobicyclic Co(III) complexes did result in some rearrangement of the ligand framework,[3] so the absence of this in the present case is noteworthy. It is possible that the observed exocyclic olefins are indeed *thermodynamically* favoured, when the consequences of intramolecular strain of the *complex* as a whole are considered, and molecular mechanics calculations are currently being pursued in order to gauge whether the observed exocyclic species are actually more stable than their endocyclic isomers.

The scope of the present reactions should, in principle, be extendable to acyclic ligands, although the stability of the pendant amine precursor complex must be such that N-nitrosation of the donor N atoms is avoided. The synthesis of compounds such as 1-6 has been motivated by a desire to extend the synthetic utility of pendant amine macrocycles toward compounds bearing other types of functional groups, as shown below, and this is currently being pursued.

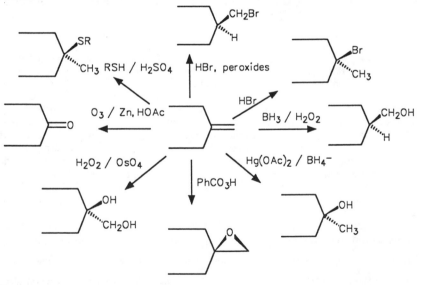

References

1.   J. March, 'Advanced Organic Chemistry', Wiley-Interscience, New York, 3rd Edn, 1985.
2.   C. J. Collins, *Acc. Chem. Res.*, 1971, **4**, 315.
3.   R. J. Geue, T. W. Hambley, J. M. Harrowfield, A. M. Sargeson and M. R. Snow, *J. Am. Chem. Soc.*, 1984, **106**, 5478.

Disilver(I) Complexes of Pendant-arm Macrocycles

Harry Adams, Neil A. Bailey, Simon R. Collinson*,
David E. Fenton, Choki Fukuhara, Paul C. Hellier,
Paul D. Hempstead, and Philip B. Leeson*

DEPARTMENT OF CHEMISTRY, THE UNIVERSITY, SHEFFIELD
S3 7HF, UK

It has been previously noted that the reaction of tris(2-ethylamino)amine (tren) with aromatic and heterocyclic dicarbonyls leads to the facile generation of Schiff base macrobicycles both via metal-templated and via non-template procedures.[1] The synthesis of the disilver(I) complex of the bibracchial tetraimine Schiff base macrocycle (1) however can be achieved via the silver(I)-templated cyclocondensation of tren with 2,6-diacetylpyridine.[2] The non-macrobicyclic nature of (1) is indicated from the i.r spectrum [3280 and 3380cm^{-1}(NH$_2$ stretching modes); 1640cm^{-1} (C=N)] and positive ion f.a.b. m.s [849 a.m.u., {(1)-BF$_4$}$^+$]

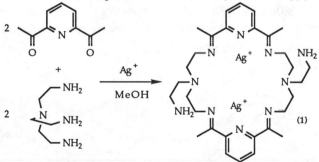

The X-ray crystal structure of the dication of (1) shows approximate C$_2$ symmetry with the two six-coordinate silver ions bound in the di-imino pyridyl head units of the macrocycle and separated by 3.17Å.[3] The two imino nitrogen atoms within each pyridine di-imine fragment show different coordination modes. One of the nitrogens acts as a bridge between the two silver ions whilst the other is bound only to a single ion. The shortest interactions shown by each silver ion are to the pyridyl nitrogen and to the primary amine on the pendant arm (2.30 to 2.36Å). There is an intermediate interaction to a non-bridging imino nitrogen donor (2.46 and 2.49Å), whilst the longest bonds are with the bridging imino nitrogens (2.68 to

2.73Å) and the tertiary amines (2.68 and 2.60Å). The macrocycle adopts a folded conformation with the pyridine groups defining a molecular cleft.

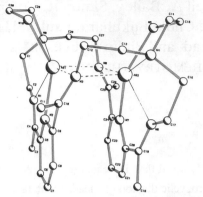

The ^{1}H and ^{13}C NMR spectral data for (1) are consistent with the macrocycle adopting a solution conformation similar to that determined in the solid state wherein the C_2 symmetry reflects the different coordination modes of the two imino nitrogen donors on each pyridine di-imine fragment. The ^{13}C NMR spectrum of (1) shows two separate signals for each carbon atom, except for those arising from the *para*-pyridine and pendant arm carbon atoms. That the macrocycle has conformationally rigid nature in the disilver complex (1) may be demonstrated by variable temperature ^{1}H NMR studies, which show the spectrum to be essentially temperature independent in the range 297 to 228K.

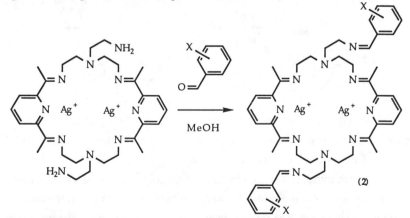

The disilver complexes derived from tren can be further functionalised by reactions with salicylaldehyde and 4-nitrobenzaldehyde to introduce Schiff bases into the pendant arms (2). The reaction of (1) with acetone in methanol gave a yellow crystalline product (3), analysis of which showed that two molecules of acetone had condensed with (1). Since the early observation of Curtis and House[4] that two molecules of acetone will condense with two primary amines, in the presence of a templating cation, to give a macrocyclic ligand, this reaction has seen wide application in macrocyclic synthesis. The spectroscopic data for (3) indicated that macrobicyclisation, as expected if the Curtis reaction pathway had been followed, had not occurred and that a bis-N-isopropylidene compound (3) had been synthesised.

(3)

The crystal structure of the dication of (3) confirms the presence of stabilised N-isopropylidene-bearing pendant arms.[5] The structure of the dication shows crystallographically imposed C_2 symmetry, with each silver atom forming five interactions with nitrogen atoms of the macrocycle (Ag-N range between 2.27 and 2.53Å); there is no silver-silver bond (Ag...Ag 5.38Å). The macrocycle has an open conformation and although somewhat folded no molecular cleft is formed and none of the nitrogen atoms form bridges between the silver atoms. In this aspect the structure is remarkably different both from that of the precursor molecule (1) and from those of related disilver(I) complexes derived from bibracchial tetraimine Schiff base macrocycles.[2]

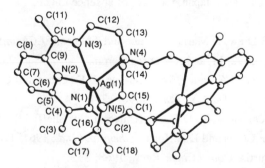

We have recently reported that in the structure of a *triangulo*-Cu₃ complex of a tetraimine Schiff base macrocycle derived from (1) via functionalisation with salicylaldehyde followed by transmetallation.[6] The *triangulo*-Cu₃ complex bears salicylideneimino-arms and the cleft-like nature of (1) is lost; a Type-3 -like pair of copper atoms bridged by a hydroxide is present and the compound has been proposed as a first generation model for the *triangulo*-Cu₃ site in ascorbate oxidase. The nature of (3) leads to the proposition that this change in molecular geometry occurs during the formation of the disilver salicylideneimino-intermediate prior to the transmetallation reaction leading to the *triangulo*-Cu₃ complex.

Reductive demetallation of (1) using NaBH₄ yields the corresponding amine macrocycle, and using H₂/Pd/C provides an unexpected route to the diprotonated cryptand (4), isolated as its tetrafluoroborate salt. Compound (4) has also been isolated from the reactions of the barium complex derived from the bibracchial tetraimine Schiff base macrocycle formed from tren and 2,6-diacetylpyridine with malonyl dichloride, and with hydrochloric acid. It therefore seems likely that the macrocyclic complexes break down under acid conditions to give their constituents which then reform to produce (4).[7]

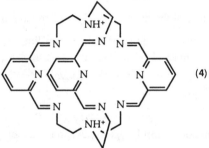

(4)

We would like to acknowledge SERC awards (to S.R.C, P.C.H and P.D.H) and the SERC and Royal Society for support and assistance in the purchase of the diffractometer. We would also like to thank the College of Science, University of the Ryukyus, Okinawa, Japan, for leave of absence to C.F.

1.   M.G.B.Drew, D.Marrs, J.Hunter and J.Nelson, *J.Chem.Soc.,Dalton Trans.*, 1992, 11, and references therein.

2.   H.Adams, N.A.Bailey, W.D.Carlisle, D.E.Fenton and G.Rossi, *J.Chem.Soc.,Dalton Trans.*, 1990, 1271.

3.   P.D.Hempstead, Ph.D.Thesis (University of Sheffield), 1991.

4.   (a) N.F.Curtis and D.A.House, *Chem and Ind.*, 1961, 1708; (b) N.F.Curtis, *Coord.Chem Revs.*, 1968, **3**, 3.

5.   H.Adams, N.A.Bailey, M.J.S.Dwyer, D.E.Fenton, P.C.Hellier and P.D.Hempstead, *J.Chem.Soc.,Chem.Commun.*, 1991, 1297.

6.   H.Adams, N.A.Bailey, D.E.Fenton, C.Fukuhara, P.C.Hellier and P.D.Hempstead, *J.Chem.Soc.,Dalton Trans.*, 1992, 729.

7.   D.E.Fenton and M.Kanesato, unpublished results, 1992.

Mercury(II)-Phenothiazine Complexes: Synthesis and Characterization

N. M. Made Gowda*, W. D. Rouch, and A. Q. Viet

DEPARTMENT OF CHEMISTRY, WESTERN ILLINOIS UNIVERSITY, MACOMB, IL 61455, USA

INTRODUCTION

Phenothiazines (PTZs) are well-known as versatile antipsychotic drugs[1]. The study of metal-phenothiazine complexes has gained importance in recent years partly due to their potential pharmacological activities. The synthesis and characterization of coordination compounds of PTZs with transition metals such as copper(II)[2], palladium(II)[3], platinum(II)[3], rhodium(II/III)[4], rhenium(VII)[5], molybdenum(IV/V)[6], and ruthenium(II)[7] have been reported. In this paper, we report the preparation and structural characterization of mercury(II) complexes of three selected PTZs (Figure 1).

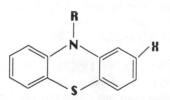

Chlorpromazine hydrochloride (CPH.Cl): R = $(CH_2)_3NHMe_2^+$ Cl$^-$; X = Cl

Promethazine hydrochloride (PMH.Cl): R = $CH_2CH(Me)NHMe_2^+$ Cl$^-$; X = H

Thioridazine hydrochloride (TRH.Cl): R = $(CH_2)_2\overline{CH(CH_2)_4NHMe}^+$ Cl$^-$; X = SMe

Figure 1   General Structure of the Phenothiazine Derivatives used

EXPERIMENTAL

Physical measurements. U.v.-vis. spectra of DMF solutions were recorded using a Perkin-Elmer model 552 spectrophotometer. I.r. spectra, both in nujol mulls and in KBr discs, were recorded using a Perkin-Elmer model 1320 i.r. spectrophotometer. Far-i.r. spectra were obtained from pressed polythene discs using a Beckman FS 720 instrument. [1]H n.m.r. spectra, recorded on a 360 MHz spectrometer, were provided by the Spectral Data Services, Inc., Champaign, IL, or recorded on a Nicolet NT-300 spectrometer. Chemical shifts were measured in p.p.m. with tetramethylsilane as the internal standard in DMSO-d6 medium. Molar conductances were determined using a conductance-resistance meter (YSI model 34, Yellow Springs, OH). Melting points (uncorrected) were measured using a Melt-Temp apparatus (Laboratory Devices, Cambridge, MA). Elemental analyses were performed by Atlantic Microlab, Inc., Norcross, GA. Mercury content in the complexes was analysed by Schwarzkopf Microanalytical lab, Woodside, N.Y.

Starting materials. HgCl2, CPH.Cl, PMH.Cl, and TRH.Cl were obtained from Aldrich and used as supplied.

General Synthetic Procedure. A solution of 6.23×10^{-3} mmol (2.00g) of PMH.Cl in minimum amount of H_2O was prepared and slowly added, with stirring, to a concentrated aqueous solution of $HgCl_2$. (6.08×10^{-3} mmol; 1.65g). The reaction mixture was magnetically stirred for 15-20 min at laboratory temperature, cooled at 4°C for 24h, and suction-filtered through an M-glass fritted funnel. The precipitate was washed with small amounts of cold H_2O and acetone, and air dried. The product was recrystallized from its saturated aqueous solution.

The same general synthetic procedure, using a metal-to-ligand mole ratio of 1:5, was used in the preparation of the complexes of CPH.Cl and TRH.Cl. The yields varied from 28 to 56%.

RESULTS AND DISCUSSION

The complexation reaction of $HgCl_2$ with the PTZ hydrochloride in aqueous medium yields a crystalline product in each case. Some physical properties and analytical data are presented in Table 1. The mononuclear 1:1 complexes obtained contain a mercury(II) center and two PTZs as principal ligands. The stoichiometric reaction scheme for the complexes is as in equation (1) where LH = CPH+, PMH+ or TRH+;

$$HgCl_2 + 2 LH.Cl \longrightarrow [Hg(LH)_2Cl_3]Cl \qquad (1)$$

The complexes of CPH.Cl and PMH.Cl are colorless while that of TRH.Cl is light yellow-green. They are non-hygroscopic and air-stable at room temperature. They do not possess sharp melting points (Table 1). The complexes are slightly soluble in water and ethanol but soluble in DMF and DMSO. The molar conductances in EtOH and DMF are in the ranges 39.7-50.8 and 71.7-103.1 ohm^{-1} cm^2 mol^{-1}, respectively, indicating an ionic ratio of 1:1.

Table 1 Physical Property and Elemental Analyses for the Complexes

Formula (F.W.)	Decomp. temp. (°C)	Elemental Data Found (Calcd.) (%)					
		C	H	N	S	Cl	Hg
[Hg(CPH)$_2$Cl$_3$]Cl (982.55)	71-82	42.7 (41.6)	4.12 (4.11)	5.90 (5.70)	-- --	21.5 (21.7)	20.0 (20.4)
[Hg(PMH)$_2$Cl$_3$]Cl (913.27)	160- 169	44.5 (44.7)	4.59 (4.64)	6.08 (6.14)	-- --	15.3 (15.5)	21.8 (22.0)
[Hg(TRH)$_2$Cl$_3$]Cl (1085.57)	86- 110	47.1 (46.5)	5.03 (5.01)	5.14 (5.16)	11.8 (11.8)	13.2 (13.1)	18.8 (18.5)

I.r. spectra. It was shown that in the case of the palladium(II) and platinum(II) complexes of PTZs the heterocyclic sulphur of the ligand was invloved in coordination as indicated by the presence of M-S stretching modes in the far-i.r. spectra of the complexes[3]. In the far-i.r. spectra of [Hg(LH)$_2$Cl$_3$]Cl there are similar bands in the ranges 300-315 cm[-1] and 320-345 cm[-1] attributable to Hg-S and Hg-Cl stretching modes, respectively. The possibility of a polymeric/oligomeric structure was ruled out based on the absence of bands below 300 cm[-1]. Furthermore, the i.r spectra of these mercury(II) complexes have shown a back-shift in the C-S bands of free PTZ ligands (670 and 730 cm[-1]) indicating involvement of the heterocyclic S atom in coordination with the metal[3,8,9].

In the free PTZ ligands, the N-H⁺......Cl interaction is observed as a broad band in the 2200-2750 cm[-1] range which either disappears or shifts with reduced intensity towards a higher wavenumber region indicating a slight weakening of the hydrogen bonding upon coordination[3,8]. This shows that the quaternary nitrogen was interacting with a chloride of the HgCl$_3$ unit through hydrogen bonding in the complex. Since there are two LH ligands involved in the complex, there should be two such hydrogen bondings present.

Electronic spectra. The u.v.-vis spectral data of [Hg(LH)$_2$Cl$_3$]Cl are as follows: λ max in nm (ε in L mol[-1] cm[-1]) 305.5 (6.433X10^3) for CPH.Cl, 297.6 (1.135X10^3) for PMH.Cl, and 306.7 (6.703X10^3) for TRH.Cl complex. There are no noticeable peaks in the visible region. Comparison of the u.v. spectra of the complexes with those of the free ligands (LH.Cl) shows a small shift in the   max . These data suggest the intraligand transitions of $\pi \rightarrow \pi^*$ type. This may be taken as an indirect evidence for involvement of the S atom in coordination[3,4].

^{1}H n.m.r spectra. The ^{1}H n.m.r. spectra of free ligands and their complexes, in DMSO-d$_6$, show complex multiplets in the ∂ 6.8-7.5 range due to aromatic protons. The chemical shift of aromatic protons is not significantly altered upon coordination. A broad singlet attributable to the side chain amine proton occurs far downfield (∂10.8 for CPH.Cl). This resonance signal shows an upfield shift indicating the weakening of the $\overset{+}{N}$-H------Cl hydrogen bonding due to complexation. The other resonance signals in the spectra, falling into the general range, ∂ 1-5, are not well resolved in some cases due to their proximity. These signals which are due to side chain (R) protons have experienced little or no change upon coordination. This may be attributed to the metal complex decomposition in DMSO, as suggested by Geary et al[3] for similar metal-PTZ complexes.

In the light of the above discussion we propose the general structure as shown in Figure 2 for the new complexes. The mercury(II) center in the complexes appears to have a square-pyramidal environment around it. The two protonated PTZ ligands coordinate to the $HgCl_3$ unit through two S atoms resulting in a square-planar $HgCl_2S_2$ unit which along with an axial Hg-Cl bond represents a square-pyramidal geometry for each complex.

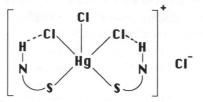

Figure 2   General structure for the complexes, $[Hg(LH)_2Cl_3]Cl$: LH = S-N-H = CPH$^+$, PMH$^+$ or TRH$^+$

REFERENCES

1.   O. Bratfos and J. O. Haug, Acta Psychiat. Scand., 1979, 60, 1; S. H. Snyder, Am. J. Psychiatry, 1976. 133, 197.
2.   B. Keshavan and R. Janardhan, Indian J. Chem., 1986, 25A, 1054.
3.   W. J. Geary, N. J. Mason, I. W. Nowell, and L. A. Nixon, J. Chem. Soc., Dalton Trans., 1982, 1103.
4.   N. M. Made Gowda, and H. P. Phyu, Transition Met. Chem., in press.
5.   N. M. Made Gowda, and H. P. Phyu, Transition Met. Chem., in press.
6.   N. M. Made Gowda, B. E. Ackerson, M. Morland, and K. S. Rangappa, Transition Met. Chem., in press.
7.   R. Kroener, M. J. Heeg, and E. Deutsch, Inorg. Chem., 1988, 27, 558.
8.   L. J. Bellamy, 'The Infrared Spectra of Complex Molecules', Methuen, London, 1964, p. 355.
9.   K. Nakamoto, 'Infrared Spectra of Inorganic and Coordination Compounds', Wiley Interscience, New York, 1970, pp.159 & 214.

Chelating Properties of Acylhydrazones in Metal Complexes. An Unexpected Copper(II)–Triazolopyridine Complex

M. Carcelli, L. Mavilla, C. Pelizzi, and G. Pelizzi

ISTITUTO DI CHIMICA GENERALE ED INORGANICA, UNIVERSITÀ DEGLI STUDI DI PARMA, VIALE DELLE SCIENZE, 43100 PARMA, ITALY

1 INTRODUCTION

As part of a more extensive study of the chelating properties of polydentate ligands in metal complexes[1,2], we have recently synthesized a new acylhydrazone of formula

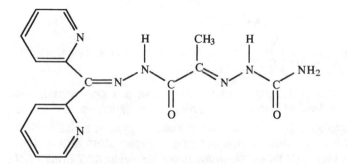

H_2dps

and are investigating its ligand behaviour towards first row transition elements.

During the reaction of H_2dps with copper(II) nitrate hydrate in ethanol solution a green powder has been isolated from which, after recrystallization from ethanol, two different copper compounds, one powdered and the other crystalline, separated. The X-ray diffraction analysis carried out on the latter revealed it contains 3-(2-pyridinyl)[1,2,3]triazolo[1,5-a]pyridine (L) as ligand to copper.

This communication describes the synthesis, the spectroscopic characterization and the X-ray structure of this complex of formula $[Cu(OH_2)_2(L)_2](NO_3)_2$.

2   EXPERIMENTAL

Di-2-pyridylketone and hydrazine hydrate are commercially available from Aldrich and Fluka, respectively.

H_2dps was obtained from a three-step process involving semicarbazide, methyl pyruvate, hydrazine and di-2-pyridylketone, following previously described procedures[3].

$[Cu(OH_2)_2(L)_2](NO_3)_2$.   To a boiling ethanol solution (100 ml) of H_2dps (0.11 g) copper nitrate hydrate (0.16 g) dissolved in ethanol (40 ml) was added. The resulting green solution (pH=4) was then refluxed for 1 h. After cooling at room temperature and by slow evaporation of the solvent a green powder was isolated.

Recrystallization from ethanol produced two green compounds, one powdered and the other crystalline (title compound). The former resulted to be a bis (pyruvate semicarbazone)copper(II) complex. The analytical and spectroscopic ($\nu(COO)_{asym} = 1635$ cm^{-1}; $\nu(COO)_{sym} = 1533$ cm^{-1}) data agree with the formula $Cu[H_2NC(O)N(H)NC(CH_3)CO_2]_2$.

3   RESULTS AND DISCUSSION

As far as the [1,2,3]triazolo[1,5-a]pyridines are concerned, the syntheses are subdivided into those starting from pyridines and those starting from triazoles[4]. The former predominate in all cases and involve the formation of a bond between the pyridine nitrogen atom and a side-chain nitrogen,   the most straightforward route involving an oxidation of the hydrazone of a pyridine-2-aldehyde or ketone[5-8].

It is remarkable that in our case the 3-(2-pyridinyl)[1,2,3]triazolo[1,5-a]pyridine derivative has been obtained in experimental conditions more milder than above.

Even if the mechanism, as a whole, is not yet well clear, the formation of the triazole system seems to be attributed to an hydrolysis process involving the H_2dps ligand upon the effect of the copper nitrate, and a successive oxidation induced by molecular oxygen, which produces the ring closure.

It can be added that the above triazolopyridine copper complex has been also obtained in good yields through a template reaction carried out in methanol solution between di-2-pyridylketone, hydrazine and copper(II) nitrate hydrate.

To the best of our knowledge, this is the first reported case of a 3-(2-pyridinyl)[1,2,3]triazolo[1,5-a]pyridine compound characterized by X-ray diffraction.

The title compound crystallizes in the monoclinic space group $P2_1/c$ with a=7.084(2), b=12.110(4), c=14.662(5) Å, β=99.10(1)°, and two molecules per unit cell. The structure was solved by conventional heavy-atom techniques and refined by least-squares methods to R and R_w of 0.0386 and 0.0406, respectively, for 1172 independent observed reflections collected by counter methods.

The structure is built up of $[Cu(OH_2)_2(L)_2]^{2+}$ cations and NO_3^- anions. The cation has a crystallographically imposed $\bar{1}$ symmetry with the copper atom located on an inversion centre. As shown in figure, the metal atom has sixfold coordination, with four close N atoms and two distant O atoms forming an elongated tetragonal octahedral environment.

The nitrate ion is disordered and the structure has been interpreted in terms of two patterns with different occupancies.

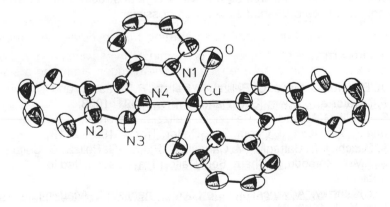

Figure   ORTEP view of the cation

Table   Bond distances (Å) and angles(°) in the coordination polyhedron

Cu - N1	2.024(3)	N1 - Cu - N4	80.1(1)
Cu - N4	2.018(3)	N1 - Cu - O	88.9(2)
Cu - O	2.400(5)	N4 - Cu - O	87.0(2)

The organic molecule, as a whole, has a near planar configuration, all its atoms lying within 0.06 Å or less from a plane. Bond distances in the triazole ring indicate a significant delocalization of π bonding and are in good agreement with those obtained for 2-phenyl-1,2,3-triazolo[4,5-b]pyrazine[9].

Packing is determined by hydrogen bonding, both the hydrogen atoms of the coordinated H_2O molecule being involved in hydrogen bonds with the oxygen atoms of the nitrate ion.

The infrared spectrum, registered in the range 4000-200 cm^{-1} using a Nicolet 5PC FT-IR spectrophotometer is characterized by the vibrational absorptions of the N=N (1037 cm^{-1}), C=N (1533 cm^{-1}) and C=C (1642 cm^{-1}) groups, the nitrate ion (1385 cm^{-1}) and the water molecule (3437 cm^{-1}). The electronic spectrum, registered in methanol solution on a Jasco 505 spectrophotometer (900-400 nm), shows an absorption at 670 nm, which is consistent with a tetragonal octahedral configuration around the copper ion[10].

For a better understanding of the mechanism of formation of the triazole-system in the copper complex, we have recently realized the synthesis of 3-(2-pyridinyl)[1,2,3]triazolo[1,5-a]pyridine from the reaction of di-2-pyridylketone and hydrazine under mild conditions. The experimental details as well as the X-ray crystal structure of the above compound will be published as soon as possible.

REFERENCES

1. A. Bonardi, C .Merlo, C. Pelizzi, G. Pelizzi, P. Tarasconi and F.Cavatorta, J. Chem. Soc. Dalton Trans., 1991, 1063.
2. S. Ianelli, G. Minardi, C. Pelizzi, G. Pelizzi,L. Reverberi, C. Solinas and P. Tarasconi, J. Chem. Soc. Dalton Trans., 1991, 2113.
3. A. Bacchi, L.P. Battaglia, M. Carcelli, C. Pelizzi, G. Pelizzi, C. Solinas and M.A. Zoroddu, J. Chem. Soc. Dalton Trans., submitted for publication.
4. A.R. Katritzky, "Advances in heterocyclic chemistry", Academic Press, New York, 1983, 34.
5. J.H. Boyer, R. Borgers and L.T. Wolford, J. Amer. Chem. Soc.,1957, 79, 678.
6. J.H. Boyer and N. Goebel, J. Org. Chem., 1960, 25, 304.
7. H. Ogura, S. Mineo, K. Nakagawa and S. Shiba, Yakugaku Zasshi, 1981, 101, 329.
8. G. Jones and D.R. Sliskovic, J. Chem. Soc. Perkin I, 1982, 967.
9. K. Yamaguchi, A. Ohsawa and C. Kawabata, Acta Cryst., 1990, C46, 1751.
10. B.J. Hathaway, J. Chem. Soc. Dalton Trans., 1972, 1196.

A Spin-frustrated Trinuclear and a Tetranuclear Imidazole-bridged Copper(II) Complex; Relevance to Multi-Copper Bio-sites

P. Chaudhuri[1,*], M. Winter[1], I. Karpenstein[1], M. Lengen[2], C. Butzlaff[2], E. Bill[2], A. Trautwein[2], U. Flörke[3], and H.-J. Haupt[3]

[1] ANORGANISCHE CHEMIE I, RUHR-UNIVERSITÄT, W-4630 BOCHUM, GERMANY
[2] INSTITUT FÜR PHYSIK, MEDIZINISCHE UNIVERSITÄT, W-2400 LÜBECK, GERMANY
[3] ALLGEMEINE ANORGANISCHE UND ANALYTISCHE CHEMIE, UNIVERSITÄT-GESAMTHOCHSCHULE, W-4790 PADERBORN, GERMANY

The trinuclear and tetranuclear copper complexes with imidazoles, an ubiquitous ligand in chemical and biological systems, are of great importance as biological models and from the magnetostructural point of view. Recently a 2.5 Å X-ray structure[1] has unambiguously demonstrated the existence of a 3+1 arrangement of the copper centers in the oxidized ascorbate oxidase, a multicopper oxidase[2] which catalyzes the four-electron reduction of dioxygen to water.

We report here a spin-frustrated triangular[3] without any μ_3-X ligand, $[L'_3Cu_3(\mu-IM)_3](ClO_4)_3$, **1**, and a tetranuclear (D_2), $[L_4Cu_4(\mu-IM)_4](ClO_4)_4 \cdot 2H_2O$, **2**, imidazolate-bridged copper(II) complex, which have been characterized by X-ray crystallography, variable temperature (2-290 K) EPR and magnetic susceptibility measurements.

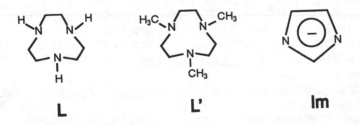

L   **L'**   **Im**

The structure of the cations in **1** and **2** are shown in Fig. 1 and 2. Each of the L'Cu (or LCu) units is coordinated via two imidazolate anions to two L'Cu (or LCu) units, yielding three in **1** and four in **2** distorted square-pyramidal CuN$_5$-polyhedra. The three copper(II) ions in **1** are arranged at the corners of an equilateral triangle with the Cu...Cu separation of 5.92 Å. The four copper ions in **2** lie on a plane to form an approximate parallelogram of sides 5.89 Å and 5.99 Å.

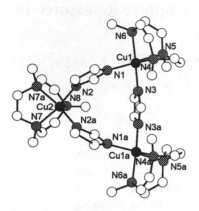

Fig.1. Structure of the trication in $\underline{\underline{1}}$

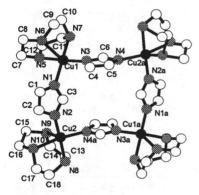

Fig.2 Structure of the cation in $\underline{\underline{2}}$

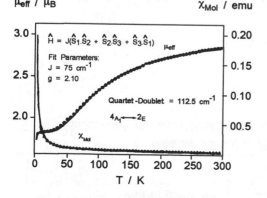

Fig. 3. Magnetic susceptibility of the trimer, 1

The molar magnetic susceptibility χ_{Mol} and the effective magnetic moment μ_{eff} of the complexes **1** and **2** are displayed in Fig. 3 and 4, which reveal a behavior typical for antiferromagnetic spin-coupling. The experimental magnetic moment for **1** decreases as the temperature is lowered until a plateau is reached at 10 K with μ_{eff} = 1.822 μ_B (Fig. 3), which is practically the theoretical value of μ_{eff} = 1.819 μ_B for an S = 1/2 ground state with g = 2.1. Below 10 K there is a decrease in μ_{eff}, reaching a value of 1.746 at 2 K. This deviation below 10 K from the theoretical value of 1.819 μ_B may be the result of saturation effect. The susceptibility data for 1 and 2 could be simulated with the spin Hamiltonians (see the inserts of Fig. 3 and 4) assuming isotropic exchange interaction between adjacent Cu(II) pairs with local spins S_i = 1/2; the simulation with the exchange coupling constants, J, are also shown as solid lines in Fig. 3 and 4.

The X-band EPR spectrum at 4.2 K recorded on a powder sample of **1** yields a quasi-isotropic g value of 2.08. The spectrum does not show any half-field (ΔM_S = 2) transition or fine structure and looks like a spectrum associated with isolated S = 1/2 spin systems possessing axial symmetry with the unpaired electron present in a $d_{x^2-y^2}$ orbital.

The X-band EPR spectra of the tetramer, **2**, recorded in the temperature range 2.7 - 270 K (Fig. 5) are governed by an isotropic resonance C1 at g 2.10. Its line width appears to be almost constant 20 - 30 mT up to room temperature, while its intensity slightly increases at higher temperatures. Above 20 K additional features of a second component C2 with wide splittings arise in the field range 80 mT to 400 mT. Its intensity exhibits significant temperature dependence with maximum at about 40 K. Analysis of the EPR spectra has led to the assignment of C2 subspectrum to the first excited triplet state |111> and the isotropic signal C1 exclusively to the highest quintet state |112> of the tetramer, **2**.

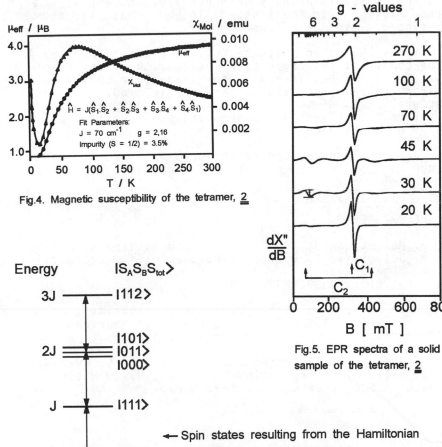

Fig.4. Magnetic susceptibility of the tetramer, **2**

$$\hat{H} = J(\hat{S}_1.\hat{S}_2 + \hat{S}_2.\hat{S}_3 + \hat{S}_3.\hat{S}_4 + \hat{S}_4.\hat{S}_1)$$

Fit Parameters:
J = 70 cm^{-1} g = 2,16
Impurity (S = 1/2) = 3.5%

Fig.5. EPR spectra of a solid sample of the tetramer, **2**

Energy $|S_A S_B S_{tot}>$

3J ——— |112>

2J ≡≡≡ |101>
        |011>
        |000>

J ——— |111>

0 ——— |110>

← Spin states resulting from the Hamiltonian (see Fig.4) for the tetramer, **2**.

The Cu-N(Im) bond distance seems to be less important in determining the magnitude of the exchange interaction than, for example, bond angles. The relevant exchange pathway through the imidazolate bridge has been described in

the literature to be only of the σ-type. Our magnetostructural analyses indicate that the σ-superexchange pathway cannot be the only mechanism for exchange coupling in
these compounds.

    The interaction between copper ions in these complexes
is much smaller than that between type III coppers in proteins which exhibit extremely strong antiparallel interactions. These observations suggest that the imidazolate-
bridged dicopper(II) structure, Cu-Im-Cu, cannot be expected to be involved in proteins having type III coppers.

    If all three metal ions in a triangle are equivalent
(eq. Cu(II)) and if the exchange interaction between the
Cu(1) and Cu(2) is antiferromagnetic in nature, then
Cu(3) senses simultaneously spin parallel and antiparallel,
thereby causing spin frustration. In our complex **1**, the
same type of frustration is present.

    It must be noted that the Cu...Cu distance of 5.92 Å
in the complex **1** is larger than the Cu...Cu distances of
3.4, 3.9 and 4.0 Å observed in the trinuclear copper unit
of ascorbate oxidase. With this caveat in mind, this new
structural class of trinuclear complexes can provide useful information toward our understanding of various properties of multicopper oxidases.

    The work is supported by the DFG. Our special thanks
are due to Prof. Dr. K. Wieghardt for his help and interest.

REFERENCES

1.   A. Messerschmidt, A. Rossi, R. Ladenstein, R. Huber,
      M. Bolognesi, G. Gatti, A. Marchesini, R. Petruzelli,
      A. Finazzi-Agro
      J.Mol.Biol. 1989, 206, 513

2.   E. I. Solomon in "Metal Clusters in Proteins", Ed.
      L. Que, Jr.; ACS Symposium Series 372; American
      Chemical Society, Washington D. C. 1988, p. 116

3.   P. Chaudhuri, I. Karpenstein, M. Winter, C. Butzlaff,
      E. Bill, A. X. Trautwein, U. Flörke, H.-J. Haupt
      J.Chem.Soc. Chem.Commun. 1992, 321.

Structural Aspects of Copper Pyrazolato Complexes

Norberto Masciocchi[1], Massimo Moret[1],
G. Attilio Ardizzoia[2], and Girolamo La Monica[2]

[1] ISTITUTO DI CHIMICA STRUTTURISTICA INORGANICA AND
[2] DIPARTIMENTO DI CHIMICA INORGANICA E CENTRO CNR.
UNIVERSITÀ DI MILANO, VIA VENEZIAN 21, 20133 MILANO, ITALY

1   INTRODUCTION

The synthesis of di- and polynuclear complexes having ligands which maintain the metal centres in close proximity is an important objective in transition metal chemistry. Interest in such systems arises because of their potential role in multi-metal centred catalysis in both biological and industrial reactions, and also because of their peculiar magnetic properties[1].

As a part of a systematic study of the chemical and structural properties of new and already known copper complexes containing pyrazoles or pyrazolato groups as ligands, we have already reported the synthesis and the characterization of the polynuclear [Cu(Hpz)$_2$Cl]$_2$,[2] [Cu(dmpz)(RNC)]$_2$,[3] [Cu$_3$(OH)(dmpz)$_3$Cl$_2$][4] and [Cu(dmpz)(OH)]$_8$[5] complexes (Hpz = pyrazole, Hdmpz = 3,5-dimethylpyrazole, R = cyclohexyl).

In the following we will briefly discuss the structural features of Cu(I) and Cu(II) complexes containing the novel dcmpz ligand (Hdcmpz = 3,5-dicarbomethoxy-pyrazole).

2   RESULTS AND DISCUSSION

The dcmpz ligand, possessing four nucleophilic centres, is, among other pyrazolato ions, highly suitable for stabilizing polynuclear systems; however, the extended π net of the heterocyclic ring and the two carboxylic branches makes the donor capability of each nucleophilic centre somewhat dependent on the electronic situation of all other atoms of the molecule; therefore, only a limited number of coordination modes is expected, as shown in Scheme I.

Starting from the [Cu(dcmpz)]$_n$ complex (1),[6] the following complexes have been prepared:

i) [Cu(dcmpz)(RNC)]$_2$ (2), by reaction with RNC in diethyl ether;
ii) Cu$_2$(dcmpz)$_2$(Py)$_2$(CO) (3), by bubbling CO in pyridine;
iii) Cu(dcmpz)(phen)(RNC) (4), *via* a multistep reaction from Cu(phen)(PPh$_3$)Cl, RCN and Hdcmpz;
iv) Cu(dcmpz)$_2$(Py)$_2$ (5), by bubbling O$_2$ in pyridine;
v) [Cu(dcmpz)$_2$]$_3$ (6), by pyridine abstraction from (5) via azeotropic distillation with benzene.

Scheme 1

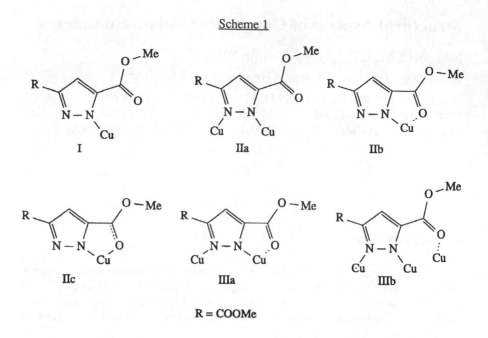

R = COOMe

The most important structural features of these compounds, collected in Table I, show that:

1) All possible coordination modes depicted in Scheme I were successfully detected.

2) The unidentate mode I, previously suggested to be present in the still elusive Cu(dcmpz)(Py)$_2$(CO) compound,[7] has been finally structurally characterized in (4), although the disorder of one of the ligands (RNC) prevented a satisfactory refinement.

3) Complex (2) possesses a Cu-[N-N]$_2$-Cu metallacycle in a boat conformation, with a dihedral angle between the two dcmpz rings of 119.6°, and wing-tip N-Cu-N angles of 97.3 and 101.4°; the same conformation is observed for compound (3), while the previously reported [Cu(dmpz)(RNC)]$_2$,[3] of imposed C$_{2h}$ symmetry, contained a strictly planar metallacycle. Under the assumption that the copper atoms must lie (more or less) in plane with the pyrazolato rings, the conformational variety of pyrazolato-bridged copper dimers is limited to boat and planar conformers. A detailed analysis of a simple geometrical model has been reported in ref.7.

Unexpectedly, the dppz derivative [Cu(dppz)(RNC)]$_2$[8] (Hdppz = 3,5-diphenylpyrazole) possesses a chair conformation, which was found only in the highly crowded environments of the [Ti(C$_5$H$_5$)$_2$(pz)]$_2$[9] and [Rh(C$_5$Me$_5$)(pz)$_3$]$_2$ dimers.[10]

In (3), a graphitic interaction between the two (parallel) pyridine ligands, which are about 3.42 Å apart, significantly decreases the Cu···Cu distance down to a value of 3.250(1) Å, which is much more similar to the values found for polynuclear Cu(II) systems containing both pyrazolato bridges and μ$_2$- or μ$_3$-OH groups[4,5,11]

4) The interaction of the -COOMe residues with the copper atoms (as low as 2.55 Å) is present only in the more acidic Cu(II) centres (complexes (5) and (6)), in apical positions with respect to the more or less distorted square-planar environments, while Cu(I)···O contacts are longer than 2.9 Å.

Table 1. Synoptic collection of the relevant geometrical parameters and i.r.absorptions [$v(CO)$] for compounds (2) to (6). Distances in Å, with e.s.d.'s in parentheses. Bold characters refer to average values. Frequencies in cm^{-1}.

Compound	(2)	(3)	(4)	(5)	(6)
Cu···Cu	3.404(1)	3.250(1)	-	-	3.500(1)
					3.480(1)
Cu - N$_{pz}$	**2.001**	**2.024**	2.01	1.950(2)	**1.986**
min.	1.992(2)	1.977(5)			1.930(3)
max.	2.010(2)	2.062(5)			2.012(3)
Cu - N^a		1.961(5)	2.04	2.057(2)	-
		2.071(5)	2.15		
Cu - C	1.849(3)^b	1.810(8)^c	1.92^b	-	-
	1.851(3)^b				
N - N	**1.350**	**1.337**	1.33	1.335(3)	**1.335**
min.	1.352(2)	1.337(6)			1.328(4)
max.	1.353(2)	1.351(6)			1.346(4)
C=O	**1.197**	**1.189**	**1.15**	**1.198**	**1.199**
min.	1.195(3)	1.175(6)	1.12	1.192(4)	1.176(6)
max.	1.206(3)	1.194(7)	1.18	1.204(3)	1.211(5)
					1.225(5)^d
					1.227(5)^d
Cu - O	-	-	-	-	2.084(3)
					2.102(3)
Cu···O	2.90	2.99	3.03	2.68	2.55
	2.90	3.00			2.69
	3.03	3.24			2.70
					2.73
Coordination mode	IIa	IIa	I	IIb	IIc,IIIa,IIIb
v(C=O)	1717	1730	1727	1712	1735
		1716	1713	1689	1721
		1705			1685
					1620^d

a) Other ligands than pyrazolate; b) RNC ligand; c) CO ligand; d) enolic COOMe residues.

5) In (6), three differently coordinated modes for the six dcmpz ligands were found; two, coordinating in the IIc fashion, show a significant elongation of the 'enolic' C=O moiety; correspondingly, the neighbouring C-C interactions shrink,[12] their double bond character being increased. Moreover, the blue species (6) is in equilibrium in CH_2Cl_2 solution with a green one (7), not containing 'enolic'-like carbonyls (i.r. evidence) which, however, could never be recovered as an analytically pure product nor crystallized. On the basis of structural analogies and spectroscopic evidence, we suggest that (7) might be a structural isomer of (6), maybe still trimeric, similar to the $[Ni(pz)_2]_3$ compound proposed by Blake *et al.*[13]

3   CONCLUSIONS

The coordinative versatility of the dcmpz ligand has been discussed on the basis of several X-ray crystal structure determinations of copper derivatives. The nature of the copper centres, as well as the presence of other ligands, determines the stoichiometry, the coordination modes and the molecular conformations of the complexes, which have been compared with their pz, dmpz and/or dppz analogues. The reader is referred to the original literature, and particularly to ref. 7 and 12, for further details.

REFERENCES

1. See for example: *Biological and Inorganic Copper Chemistry*, K.D.Karlin and J.Zubieta Eds., Adenine Press, New York, **1986**.
2. M.A.Angaroni, G.A.Ardizzoia, T.Beringhelli, G.D'Alfonso, G.La Monica, N.Masciocchi and M.Moret, *J.Organomet.Chem.*, **1989**, *363*, 409.
3. G.A.Ardizzoia, M.A.Angaroni, G.La Monica, N.Masciocchi and M.Moret, *J.Chem.Soc.,Dalton Trans.*, **1990**, 2277.
4. M.A.Angaroni, G.A.Ardizzoia, T.Beringhelli, G.La Monica, D.Gatteschi, N.Masciocchi and M.Moret, *J.Chem.Soc.,Dalton Trans.*, **1990**, 3305.
5. G.A.Ardizzoia, M.A.Angaroni, G.La Monica, F.Cariati, S.Cenini. M.Moret and N.Masciocchi, *Inorg.Chem.*, **1991**, *30*, 4347.
6. G.A.Ardizzoia and G.La Monica, *Inorg.Synth.*, *submitted*.
7. G.A.Ardizzoia, E.M.Beccalli, G.La Monica, N.Masciocchi and M.Moret, *Inorg.Chem.*, **1992**, in press.
8. G.A.Ardizzoia, G.La Monica, N.Masciocchi and M.Moret, *unpublished results*.
9. B.F.Fieselman and G.D.Stucky, *Inorg.Chem.*, **1978**, *17*, 2074.
10. L.A.Oro, D.Carmona, M.P.Lamata, C.Foces-Foces and F.H.Cano, *Inorg.Chim.Acta*, **1985**, *97*, 19.
11. F.B.Hulsbergen, R.W.M. ten Hoedt, G.C.Verschoor, J.Reedijk and A.Spek, *J.Chem.Soc.,Dalton Trans.*, **1983**, 539.
12. M.A.Angaroni, G.A.Ardizzoia, G.La Monica, E.M.Beccalli, N.Masciocchi and M.Moret, *J.Chem.Soc.,Dalton Trans.*, **1992**, in press.
13. A.B.Blake, D.F.Ewing, J.E.Hamlin and J.M.Lockyer, *J.Chem.Soc.,Dalton Trans.*, **1977**, 1897.

Low-coordinate Chalcogenolato Complexes of Zinc, Cadmium, and Mercury

Manfred Bochmann*, Gabriel C. Bwembya, Kevin J. Webb, and R. Grinter

SCHOOL OF CHEMICAL SCIENCES, UNIVERSITY OF EAST ANGLIA, NORWICH NR4 7TJ, UK

Neutral chalcogenolato complexes of divalent metals such as zinc and cadmium are generally known to adopt structures in which the chalcogenolato ligands form bridges between tetrahedrally coordinated metal ions to give two- or three-dimensional infinite lattices.[1] The degree of association and the extent of cross-linking is, however, sensitive to the steric requirements of the chalcogenolato ligands, and with suitably bulky ligands the formation of coordination polymers can be prevented.[2] As part of our search for single-source precursors for II-VI solid-state materials[2-4] we have investigated the coordination chemistry of sterically highly hindered ligands such as $2,4,6\text{-}But_3C_6H_2EH$ (Ar"EH; E = S, Se, Te), as well as of their methyl and isopropyl-substituted analogues.[5,6]

The preferred route to sterically hindered chalcogenolato complexes is the protolysis of metalbis(amide)s with chalcogenols in hexane (eq. 1); this method avoids the possibility of ionic impurities. The products are obtained as moderately hexane soluble precipitates which are recrystallised from toluene and sublime at elevated temperatures (150 - 230°C/0.01 torr).

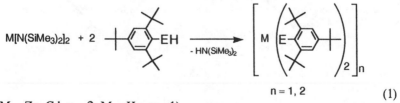

$$n = 1, 2 \tag{1}$$

(M = Zn, Cd, n = 2; M = Hg, n = 1)

Crystals of $Zn(SAr")_2$ suitable for X-ray diffraction were grown by sublimation at 150°C/0.01 torr (Fig. 1).[7] Its structure resembles that of the cadmium thiolato and selenolato complexes.[2,4] These compounds are the first examples of molecular, non-polymeric zinc and cadmium chalcogenolates. All three complexes are dimeric, with trigonal-planar three-coordinate metal centres. Although in view of the differing atomic radii of Zn, Cd, S and Se the degree of steric hindrance in these complexes must vary significantly, a comparison of all three structures shows a number of distinctive similarities:

(a) the terminal metal-ligand bonds are significantly shorter (ca. 0.1 Å) than bridging M-E distances;
(b) the chalcogenolato bridges in the M_2E_2 ring are asymmetric;
(c) the substituents on the bridging ligands are moved out of the yz plane of the molecule.

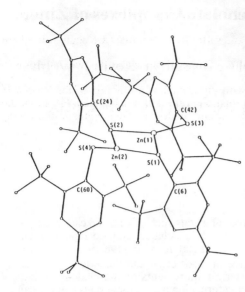

Figure 1. Molecular structure of [Zn(SAr")₂]₂. Selected bond lengths and angles: Zn(1)-S(1) 2.371(5); Zn(1)-S(2) 2.285(5); Zn(1)-S(3) 2.184(5) S(1)-Zn(2) 2.277(5); S(2)-Zn(2) 2.429(5); S(4)-Zn(2) 2.203(5) Å; S(2)-Zn(1)-S(1) 84.5(2)°; S(3)-Zn(1)-S(1) 120.8(2)°; S(3)-Zn(1)-S(2) 154.2(1)°; C(6)-S(1)-Zn(1) 130.6(2)°.

The bond length distribution in dimeric zinc and cadmium complexes is illustrated schematically below. The shorter bond lengths are indicated by bold lines. In all structures the shorter bridging metal-ligand distances are associated with the more acute of the R-E-M angles.

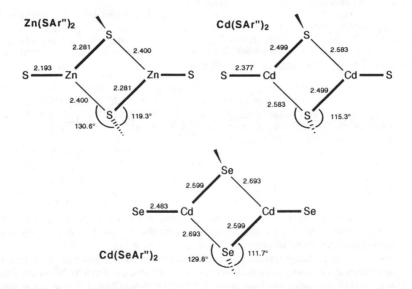

In the absence of convincing steric arguments the observed distortion of the M_2E_2 ring is likely to be of electronic origin. This was confirmed by EHMO calculations on a simplified model where the aryl substituents are replaced by hydrogen atoms placed close to the crystallographically determined *ipso*-carbon positions. The Zn - S bond length was set to an average value of 2.29 Å. With such constraints the overlap populations are a qualitative measure of the relative bond strengths, with higher values indicating more efficient overlap (Figure 2). The results indicate:

(1) the terminal Zn-S bonds are strong due to efficient overlap between Zn s and p_x orbitals with sulfur s and p_x;

(2) bonding contributions from zinc and sulfur p_z or d-orbitals are negligible (there is essentially no p_π-p_π interaction);

(2) overlap between the bridging sulfur s and p_x orbitals and the two Zn atoms is almost identical, but

(3) overlap of the sulfur p_y orbital with Zn orbitals is a function of the Zn-S$_{br}$-H angle. This is illustrated in Fig 3: on rotation of the sulfur substituent H1 about the S(1)-S(2) axis the position of the shorter and the longer Zn-S bonds in the Zn_2S_2 ring are interchanged.

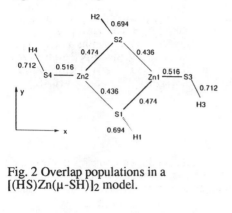

Fig. 2 Overlap populations in a $[(HS)Zn(\mu-SH)]_2$ model.

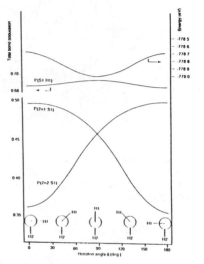

Fig. 3 Dependence of the populations of the Zn1-S1 and Zn2-S1 bonds on the angle of rotation of H1 with respect to the S1-S2 axis.

In summary, while the origin of the displacement of the substituents of the bridging ligands out of the molecular yz plane (and away from the energy minimum) is most likely steric, this induces a series of changes in orbital overlaps within the M_2E_2 skeleton which lead to the formation of asymmetric chalcogenide bridges and the observed bond length distribution.

The dimeric Zn and Cd complexes dissociate extensively in non-polar solvents to give two-coordinate monomers, as indicated by molecular weight determinations in benzene.[7] As expected for coordinatively unsaturated complexes $Zn(SAr'')_2$ reacts readily with N, O or P-donor ligands in hexane to give three-coordinated adducts, such as $Zn(SAr'')_2(L)$ (L = 2,6-lutidine, NC_5H_4CHO, THF, SC_4H_8, PhCN). With N-methylimidazole a four-coordinate complex is formed which loses one nitrogen ligand reversibly to give $Zn(SAr'')_2(imid)$ on recrystallisation from toluene.

Few three-coordinate chalcogenolato complexes of Zn and Cd are known.[8] The structures of $Zn(SAr'')_2(2,6$-lutidine) (Fig. 4) and of $Zn(SAr'')_2(imid)_2$ have be determined and may be compared to $Hg(SAr'')_2(pyridine)$.[9] The main structural features are represented on Fig. 5.

The comparison of the three- and the four-coordinate Zn complexes demonstrates that, in spite of the bulkiness of the SAr'' ligands, steric hindrance is not an important factor in determining bond lengths: surprisingly, the Zn-N distance in tetrahedral $Zn(SAr'')_2(imid)_2$ is ca. 0.1 Å shorter than in the 3-coordinate $Zn(SAr'')_2(lutidine)$. Rather, the metal-ligand bond lengths are a function of the S-Zn-S angle and appear to correlate with the degree of s-character of the metal orbitals of the $Zn(SR)_2$ fragment involved in ligand bonding. Whereas Zn in $Zn(SAr'')_2(imid)_2$ is approximately sp^3-hybridised, the wide S-Zn-S angle in Zn(SAr")$_2$(lutidine) suggests sp hybridisation, with a more loosely associated donor ligand. This bonding pattern is further illustrated by Hg(SAr")$_2$(py) where the S-Hg-S angle deviates little from 180°, with a very long Hg-N distance of 2.68 Å. The results caution against a facile correlation between bond length and bond order unless the nature of the bonding orbitals is also known and comparable.

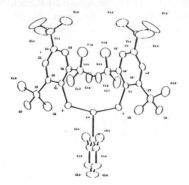

Figure 4. Molecular structure of Zn(SAr")$_2$(NC$_5$H$_3$Me$_2$-2,6). Selected bond lengths and angles: Zn-S 2.207(3); Zn-N 2.144(7) Å; S-Zn-S' 156.26(4)°; N-Zn-S 101.87(4)°; C(1)-S-Zn 110.4(2)°.

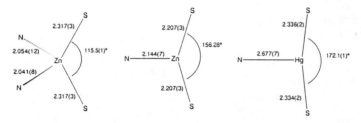

Figure 5. Comparison of the main structural parameters of M(SR)$_2$(L)$_n$ complexes.

Acknowledgements. This work is supported by the Science and Engineering Research Council (21st Century Materials Initiative). We are grateful to Professor M.B. Hursthouse and Dr. M. Mazid (University College Cardiff) for the crystal structure determinations.

References
1. I.G. Dance, *Polyhedron* 1986, **5**, 1037; P.J. Blower and J.R. Dilworth, *Coord. Chem. Rev.* 1987, **76**, 121;
2. M. Bochmann, K.J. Webb, M. Harman, and M.B. Hursthouse, *Angew. Chem.* 1990, **102**, 703; *Angew. Chem. Int. Ed. Engl.* 1990, **29**, 638.
3. M. Bochmann and K.J. Webb, *Mat. Res. Soc. Symp. Proc.* 1991, **204**, 149; M. Bochmann, K.J. Webb, J.E. Hails and D. Wolverson, *Eur. J. Solid State Inorg. Chem.*, in press.
4. M. Bochmann, K.J. Webb, M.B. Hursthouse and M. Mazid, *J. Chem. Soc., Dalton Trans.* 1991, 2317.
5. M. Bochmann and K.J. Webb, *J. Chem. Soc., Dalton Trans.* 1991, 2325; M. Bochmann, A.P. Coleman and A.K. Powell, *Polyhedron* 1992, **11**, 507.
6. M. Bochmann, K.J. Webb, A.P. Coleman, M.B. Hursthouse and M. Mazid, *Angew. Chem.* 1991, **103**, 975; *Angew. Chem. Int. Ed. Engl.* 1991, **30**, 973.
7. M. Bochmann, G. Bwembya, R. Grinter, J. Lu, K.J. Webb, D.J. Williamson, M.B. Hursthouse and M. Mazid, *Inorg. Chem.*, submitted.
8. E.S. Gruff and S.A. Koch, *J. Am. Chem. Soc.* 1989, **111**, 8762; E.S. Gruff and S.A. Koch, *J. Am. Chem. Soc.* 1990, **112**, 1245; P.P. Power and S.C. Shoner, *Angew. Chem.* 1990, **102**, 1484; *Angew. Chem. Int. Ed. Engl.* 1990, **29**, 1403; W. Wojnowski, B. Becker, L. Walz, K. Peters, E.M. Peters and H.G. von Schnering, *Polyhedron* 1992, **11**, 607.
9. M. Bochmann, K.J. Webb and A.K. Powell, *Polyhedron* 1992, **11**, 513.

Cyclohexanethiolato Complexes of Mercury

W. Clegg[1], K. A. Fraser[1], I. C. Taylor[1], T. Alsina[2], and J. Sola[2]

1 DEPARTMENT OF CHEMISTRY, UNIVERSITY OF NEWCASTLE UPON TYNE, NEWCASTLE UPON TYNE NE1 7RU, UK

[2] DEPARTAMENT DE QUÍMICA, UNIVERSITAT AUTÒNOMA DE BARCELONA, 08193 BELLATERRA, BARCELONA, SPAIN

1. INTRODUCTION

Thiolate and related complexes play an important part in the biological chemistry of Zn, Cd and Hg.[1,2] The particular affinity of thiols for mercury is well expressed in the alternative name of mercaptans for this class of compounds.

In the coordination chemistry of thiolate complexes, tetrahedral coordination predominates for Zn and Cd, but linear and trigonal-planar coordination are also important for Hg.[3] All three geometries have been proposed for Hg atoms in biological systems.[2,4] It is far from clear what factors determine the coordination of Hg in thiolate or other complexes, and a wide variety of structural types has previously been found.

Benzenethiolate has been used frequently in studies of metal thiolate complexes. We show here some structural results for the cyclohexanethiolate ligand. This was chosen in order to compare ultra-violet spectroscopic data with those available for cysteine complexes of biological importance, benzenethiolate giving other bands which obscure those of interest. The complexes described are polymeric $[Hg(SR)_2]_n$ (1), $[Hg(SR)_3]^-$ obtained as its $[Et_4N]^+$ salt (2), polymeric $[Hg(SR)I]_n$ (3), and the homologous series $[Hg_7(SR)_{12}X_2]$ (X=Cl, 4; X=Br, 5; X=I, 6); R is cyclo-C_6H_{11} in each case.

2. EXPERIMENTAL

Complex 1 was obtained from a 2:1 stoichiometric reaction of NaSR with $HgCl_2$ in acetonitrile. A 3:1 reaction, followed by addition of Et_4NCl, produced the salt 2. Initial attempts to prepare [Hg(SR)Br], the bromide analogue of 3, gave instead the unexpected complex 5, and the analogous complexes 4 and 6 were subsequently prepared in a similar way. Complex 3 was eventually obtained from a very slow diffusion of solutions of HgI_2 and RSH.

Satisfactory chemical analyses were obtained in all cases. Crystal structures were determined from data collected on a Stoe-Siemens diffractometer with MoKα radiation, corrected for absorption effects. For complexes 4, 5 and 6, low-temperature data collection was essential for a satisfactory refinement of the structures, but the temperature could not be taken below 200K, because of deterioration of the crystals on further cooling, possibly involving the onset of structural phase transitions.

3. RESULTS

Complex 1

Among about a dozen known structures of complexes of formula $Hg(SR)_2$, this stands out as unique. Most other structures have either essentially linear two-fold coordination of Hg in discrete monomeric molecules, with much weaker Hg...S intermolecular interactions, or distorted tetrahedral coordination of Hg by approximately symmetrical sulfur bridges in a linear polymeric chain.[5] Complex **1** is intermediate between these forms. Each mercury atom forms two primary Hg—S bonds in a markedly bent HgS_2 unit [Hg—S 2.372 and 2.374Å, S—Hg—S 160.4°] which forms two secondary Hg—S bonds (2.959 and 3.004Å) with adjacent units, to give a polymeric chain with irregular, highly distorted, tetrahedral Hg coordination and very asymmetric sulfur bridges (Figure 1).

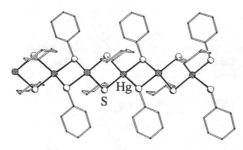

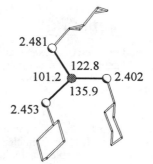

Figure 1 Chain polymeric structure of **1**

Figure 2 Structure of the anion of **2**, with bonds (Å) and angles (°) at Hg

Complex 2

This contains discrete cations and anions, with no secondary Hg...S interactions. Coordination of Hg is trigonal planar with considerable in-plane distortions (Figure 2). The longest Hg—S bond lies opposite the largest angle, and the shortest bond lies opposite the smallest angle. This pattern is well established for three-coordinate d^{10} complexes with various metals and ligands. It can be considered as a tendency towards two-coordination, with lengthening of one bond and opening up of the opposite angle.[6]

Complex 3

This has a polymeric sheet structure (Figure 3). Chains of alternating Hg and S atoms are linked together by iodide anions. Each sulfur atom bridges two mercury atoms, while the iodides are triply bridging with one short (2.772Å) and two long (3.384 and 3.529Å) Hg—I bonds. The sulfur bridges are somewhat asymmetric (Hg—S 2.384 and 2.589Å).

Of particular interest is the presence of two quite different coordination environments for mercury: half the Hg atoms are approximately tetrahedrally coordinated (by two S and two I, with angles between 102.1 and 118.3°), while the other half show a distorted octahedral coordination (by four I and two *trans* S, the S—Hg—S angle being 167.4°). The octahedrally coordinated Hg atoms have somewhat shorter bonds to S (2.384 versus 3.529Å), but much longer bonds to I (3.384 and 3.529 versus 2.772Å), so they can be regarded as primary linear HgS_2 units with additional weak Hg...I interactions approximately normal to the primary bonds, similar to the mercury coordination in complex **1**.

This structure is very similar to that observed for Hg(SPr^i)Cl.[7] In the structures of Hg(SMe)Cl and Hg(SMe)Br, polymeric sheets are also observed, with triply bridging halides and doubly bridging thiolates.[8] By contrast, however, only pseudo-octahedral coordination of Hg is seen in these two complexes (two bonds to S and three to Cl or Br, with a weak secondary Hg...S interaction in the sixth coordination site).

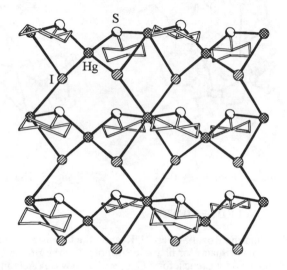

Figure 3 Sheet polymeric structure of **3**

Complexes 4, 5 and 6

These discrete molecules (Figure 4), which all crystallize isostructurally with exact three-fold rotation symmetry in a rhombohedral space group, display a number of unusual features.

A central X (Cl, Br or I) atom is surrounded by a distorted octahedral array of Hg atoms; the other X atom, in marked contrast, is terminally bonded to the seventh Hg atom, with a much shorter Hg—X bond. The twelve S atoms of the thiolate ligands form a relatively undistorted octahedron surrounding the central XHg$_6$ core (Figure 5). The six central Hg atoms have a highly distorted four-fold coordination geometry with two short (2.287 to 2.423Å) and one long (2.903 to 3.244Å) Hg—X distance. They can best be considered as approximately linear S—Hg—S units with secondary Hg...S and Hg...X interactions normal to the primary bonds and responsible for holding the molecular cluster together. The seventh Hg atom has almost regular tetrahedral coordination geometry. Three thiolates are terminally bonded; the other nine act as asymmetric bridges.

Across the series from complex **4** to **5** and to **6**, as the halogen atom increases in size, there is an increase in the Hg—X bond lengths, as expected. Differences in the thiolate bonding, however, appear to be unimportant, with no clear trends in distances or angles.

The variety of Hg coordination geometries found within these molecules, from essentially regular tetrahedral through an intermediate two/three-coordination by thiolates to almost linear two-coordination, may be related to the metal sites in Hg-metallothioneins; incorporation of more than four Hg(II) ions probably leads to a progressive change from tetrahedral to essentially linear coordination.[4]

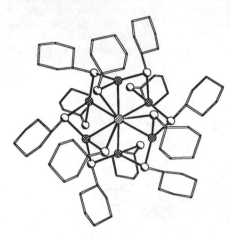

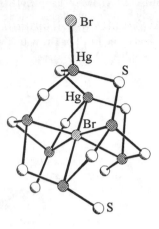

Figure 4 Structure of complexes 4—6   Figure 5 The central core

4. ACKNOWLEDGEMENTS

We thank SERC and the Royal Society (UK), and the Comisión Interministerial de Ciencia y Technología (Spain) for financial support; and the British Council and the Ministerio de Educación y Ciencia for a Cooperative Award (Acción Integrada).

5. REFERENCES

1.   'Metallothioneins II', eds. J.H.R. Kägi and Y. Kojima, Birkhäuser, Basle, 1987; W.F. Furey, A.H. Robbins, L.L. Clancy, D.R. Winge, B.C. Wang and C.D. Stout, Science, 1986, 231, 704; E.S. Gruff and S.A. Koch, J.Am.Chem.Soc., 1990, 112, 1245.
2.   J.G. Wright, H. Tsang, J.E. Penner-Hahn and T.V. O'Halloran, J.Am.Chem.Soc., 1990, 112, 2434; S.P. Watton, J.G. Wright, F.M. MacDonnell, J.W. Bryson, M. Sabat and T.V. O'Halloran, J.Am.Chem.Soc., 1990, 112, 2824; J.G. Wright, M.J. Natan, F.M. MacDonnell, D.M. Ralston and T.V. O'Halloran, Prog.Inorg.Chem., 1990, 38, 323.
3.   I.G. Dance, Polyhedron, 1986, 5, 1037; P.J. Blower and J.R. Dilworth, Coord.Chem.Rev., 1987, 76, 121.
4.   B.A. Johnson and I.M. Armitage, Inorg.Chem., 1987, 26, 3139.
5.   For comprehensive references, see T. Alsina, W. Clegg, K.A. Fraser and J.Sola, J.Chem.Soc.,Dalton Trans., 1992, 1393.
6.   R.A. Santos, E.S. Gruff, S.A.Koch and G.S. Harbison, J.Am.Chem.Soc., 1991, 113, 469.
7.   P. Biscarini, E. Foresti and G. Pradella, J.Chem.Soc.,Dalton Trans., 1984, 953.
8.   A.J. Canty, C.L. Raston and A.H. White, Aust.J.Chem., 1979, 32, 311 and 1165.

Mercury and Methylmercury Complexes with the Tripod Ligand $N(CH_2CH_2PPh_2)_3$

Carlo A. Ghilardi, Stefano Midollini, and Annabella Orlandini

ISTITUTO PER LO STUDIO DELLA STEREOCHIMICA ED ENERGETICA DEI COMPOSTI DI COORDINAZIONE, CNR, VIA J. NARDI, 39, 50132 FIRENZE, ITALY

Mercury and methylmercury complexes have been largely investigated owing to the growing concern about the complexation of mercury in the environment.[1] The $HgCH_3^+$ ion has a strong tendency toward a final linear coordination, even if in some complexes, secondary (and weaker) interactions have been ascertained, mainly in the solid state.[2] Renewed interest in coordinations >2, in methylmercury complexes, stems from the nature's mercury detoxification catalysts, which have been recently investigated.[3]

We report the complexation of the Hg^{2+} and $HgCH_3^+$ ions by the tripod-like ligand $N(CH_2CH_2PPh_2)_3$, np3. The latter ligand, due its flexibility, is known to stabilize several geometries, acting as bi, tri, or tetradentate.[4]

The reactions of np3 with mercury halides in presence of ($\underline{n}$-Bu4N)PF6 or methylmercury triflate, in organic solutions, allow the clean isolation of the complexes [(np3)HgX]PF6 (X = Cl, Br, I) and [(np3)HgCH3]CF3SO3.

An X-ray diffraction study has established that in the methyl- as well in the iodide-derivative the metal is tetrahedrally coordinated, the np3 acting as tridentate ligand: as a matter of fact the central nitrogen atom is 3.50 and 3.07 Å from the mercury respectively. The methylmercury derivative represents a practically unique example of an organomercury complex with the metal in tetrahedral environment. The methyl derivative displays Hg-P distances significantly longer than those in the iodide derivative; moreover one Hg-P distance is longer than the other two (Table 1).

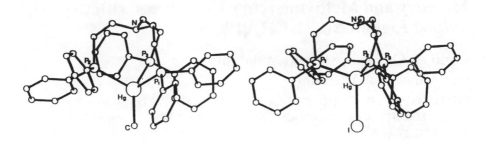

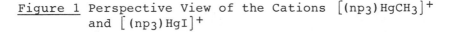

<u>Figure 1</u> Perspective View of the Cations $[(np_3)HgCH_3]^+$
          and $[(np_3)HgI]^+$

The ^{31}P and ^{199}Hg spectra of the halide-mercury
complexes in CH_2Cl_2 solutions are in agreement with the
structural results, the NMR parameters being consistent
with the literature data[5] (i.e. $[(np_3)HgI]^+$:$^{31}P\{^1H\}$,
δ = -5.3 ppm (s with ^{199}Hg satellites, J(HgP) = 2700 Hz;
$^{199}Hg\{^1H\}$, δ = 1732 ppm (q)). In spite of the X-ray
structural analogy between the methyl and iodide complexes, only the former shows in solution a complicated
dynamic behaviour, depending on the temperature and the
concentration. In the overall temperature range so far
investigated the ^{31}P spectra (and the corresponding
^{199}Hg ones) show only one signal, all of the three
P atoms of np_3 appearing co-ordinated and equivalent
(Figure 2). Quite interestingly, the value of the $^1J(HgP)$
coupling constant, which is exceptionally small,
increases continuously as the temperature drops; this
value also increases with the solution concentration
(Table 2). At temperatures >ca. 230K the Hg-P spin
correlation is lost so that a broad singlet is observed
at room temperature. This last finding is consistent
with a rapid ligand dissociation and it has been
confirmed by recording the spectra in presence of

<u>Table 1</u> Selected Bond Distances and Angles for
         $[(np_3)HgX]^+$

	X=CH$_3$	X=I		X=CH$_3$	X=I
Hg-P1	2.600(8)	2.548(8)	Hg-X	2.18(3)	2.800(2)
Hg-P2	2.808(7)	2.529(8)	Hg---N	3.50(2)	3.07(2)
Hg-P3	2.615(9)	2.523(9)			
P1-Hg-P2	100.6(3)	108.7(3)	P1-Hg-X	121.6(9)	107.4(2)
P1-Hg-P3	100.7(3)	107.7(3)	P2-Hg-X	111.0(9)	108.8(2)
P2-Hg-P3	97.1(3)	109.9(3)	P3-Hg-X	121.5(9)	114.2(2)

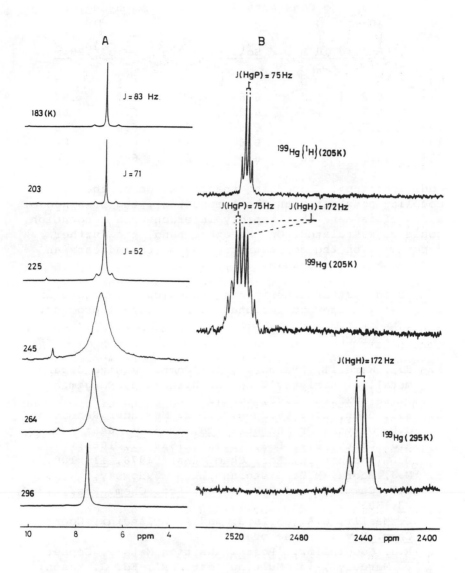

<u>Figure 2</u> Variable Temperature ^{31}P and ^{199}Hg NMR Spectra of CH$_2$Cl$_2$ Solutions of [(np$_3$)HgCH$_3$]CF$_3$SO$_3$. A: ^{31}P{^{1}H}, Sol. 0.15 M; B: ^{199}Hg and ^{199}Hg{^{1}H}, Sol. 0.20 M (Chemical Shifts are relative to external H$_3$PO$_4$ (^{31}P) and 0.1 M Hg(ClO$_4$)$_2$ in 0.1 M HClO$_4$ (^{199}Hg))

Table 2 ^{31}P Chemical Shifts (δ, ppm) and $^1J(HgP)$
Coupling Constants (Hz) for $[(np_3)HgCH_3]CF_3SO_3$
in CH_2Cl_2

	Sol. 0.10 M		Sol. 0.15 M		Sol. 0.20 M	
TK	δ	J	δ	J	δ	J
225	6.8	47	6.8	52	6.8	57
215	6.7	47	6.8	62	6.8	66
203	6.7	66	6.8	71	6.8	75
193	6.7	73	6.8	77	6.8	82
183	6.7	80	6.8	83	6.8	90

additional np_3. Concerning the variation of the
coupling constant with the temperature, the presence of
a set of isomers that rapidly interconverts in solution,
could be postulated. On the other hand, the further
dependence on the concentration seems to indicate an
intramolecular exchange.

Both further experiments and calculations are in
progress in order to rationalize the overall process.

REFERENCES

1. J.L. Wardell, 'Mercury', in 'Comprehensive Organo-
    metallic Chemistry', Ed. G. Wilkinson, Pergamon
    Press, Oxford, 1982, Vol. 2.
2. R.D. Bach, H.B. Vardhan, A.F.M. Maqsudur Rahman and
    J.P. Oliver, Organometallics, 1985, 4, 846;
    A.J. Canty, C.V. Lee, ibid, 1982, 1, 1063;
    D.L. Rabenstein, Acc. Chem. Res., 1978, 11, 100.
3. M.J. Moore, M.D. Distefano, L.D. Zydowsky,
    R.T. Cummings and C.T. Walsh, Acc. Chem. Res., 1990,
    23, 301.
4. C. Mealli, C.A. Ghilardi and A. Orlandini, Coord.
    Chem. Rev., in the press.
5. R.J. Goodfellow, 'Post-Transition Metals, Copper
    to Mercury', in 'Multinuclear NMR', Ed. J. Mason,
    Plenum Press, New York and London, 1987.

Preparation and Uses of New Mercury, Gold, and Silver Compounds

J. Vicente[1], M. D. Bermúdez[2], F. J. Carrión[2],
and G. Martínez Nicolás[2]

[1] DEPARTAMENTO DE QUÍMICA INORGÁNICA, GRUPO DE QUÍMICA ORGANOMETÁLICA, UNIVERSIDAD DE MURCIA, 30171-ESPINARDO, MURCIA, SPAIN

[2] DEPARTAMENTO DE QUÍMICA APLICADA E IGENIERÍA METALÚRGICA, GRUPO DE CIENCIA DE MATERIALES E INGENIERÍA METALÚRGICA, UNIVERSIDAD DE MURCIA, 30203-CARTAGENA, SPAIN

1. INTRODUCTION.

In our interest to develop the chemistry of orthometallatedgold(III) complexes we have prepared new species by transmetallation processes using organomercury complexes. In this way we have obtained mono- di- or triarylgold(III) derivatives by one or two successive transmetallation reactions. The study of the reactivity of these new species has led to some unprecedented results such as the new acetylacetonategold(III) complexes where the diketone is acting as chelate through both oxygen atoms, and the synthesis of cyclopentadienylgold(III) species in a simple way. We also describe the reactivity of the arylketonyl species to give C-C coupling by reductive elimination processes.

2. NEW ARYLGOLD(III) COMPLEXES.

When azobenzenemercury derivatives substituted by long alkyl chains $HgC_6H_3(N=NC_6H_4C_{10}H_{21}-4')-2$, $C_{10}H_{21}-5Cl$ obtained by modified lit. methods, are used as transmetallating reagents, results are somewhat different to that already described by us[2]. Thus, anionic trichloroaryl species have been prepared now that were found to exist only in solution when 2-(phenylazo)phenyl derivatives are used. Although the free substituted azobenzene ligand is a liquid crystal, no mesomorphic behaviour has been found for the o-metallated compounds.

In the development of the organomercury route to new gold compounds, we have prepared the first neutral triarylgold(III) species by successive transmetallation reactions that led to complexes with one chelating ligand and two monocoordinate o-substituted phenyls such as o-nitro or o-trifluoromethylphenyl groups.

The synthesis of bis(o-nitrophenyl)gold derivatives has been described earlier by use of the diarylmercury derivative as transmetallating reagent[3]; we now describe the synthesis of the same species by a more direct method by symmetrization of the chloarylmercury compound. The high stability of the bis(o-nitrophenyl)gold complexes has allowed us to prepare a bis(o-nitrophenyl)acetylacetonate (acac) derivative where the acac ligand is acting as a chelate bonding to the gold atom through both oxygen atoms. We are currently studying the reactivity of this new species towards protonated ligands.

Cyclopentadienylgold(I) complexes have been known for some time[4], but to the best of our knowledge, no cyclopentadienylgold(III) species have been reported. We have prepared arylcyclopentadienylgold(III) complexes in a simple way by reacting aryldichlorocomplexes (aryl= 2-phenylazophenyl or 2-(dimethylaminomethyl)phenyl) with thalliumcyclopentadienyl. The σ-cyclopentadienyl group appears as a unique resonance in the ^{1}H n.m.r. at room temperature. Variable temperature n.m.r. studies must be carried out to establish the fluxional behaviour of these species.

We have described the synthesis of ketonylgold(III) derivatives by C-H activation of methylketones.[5] We are currently studying the use of these new species as intermediates in C-C bond formation by coupling of both mutually cis organic ligands in a reductive elimination process similar to that already described for cis-diarylgold(III) complexes.[6] So far we have been able to isolate and spectroscopically characterize the products of condensation of the acetonyl group with C6H4N=NPh and C6H3(N=NC6H4Me-4')-2, Me-5 groups.

Finally, we have prepared a Schiff base by condensation of the chiral amine 1-phenylethylamine with decanal. Attempts to mercuriate this imine have failed, but we have obtained silver adducts and are currently studying their use as precursors of new gold compounds.

REFERENCES

1. A.P. Terent'ev and Y.D. Mogilyanskii, Zhur. Obschei Khim.,1958, 28, 1959.(C.A.1958, 1327b); V.I. Sokolov, L.L. Troitsksya and O.A. Reutov. J. Organomet. Chem., 1975, 93, C11.

2. J. Vicente, M.T. Chicote and M.D. Bermúdez. Inorg. Chim. Acta, 1982, 63, 35.

3. J. Vicente, M.T. Chicote, A. Arcas y M. Artigao. Inorg. Chim. Acta, 1982, 65, L251.

4. H. Werner, H. Otto, T. Ngo-Khac and C. Burschka, J. Organomet. Chem., 1984, 262, 123.

5. J. Vicente, M.D. Bermúdez, M.T. Chicote and M.J. Sánchez J. Chem. Soc. Chem Commun. 1989, 141; J. Chem. Soc. Dalton Trans. 1990, 1945.
6. J. Vicente, M. D. Bermúdez and J. Escribano. Organometallics, 1991, 10.

Bis[bis(trimethylsilyl)methyl]zinc – a Valuable Synthon to Study the Formation of Zincates

Matthias Westerhausen* and Bernd Rademacher

INSTITUT FÜR ANORGANISCHE CHEMIE DER UNIVERSITÄT STUTTGART, PFAFFENWALDRING 55, W-7000 STUTTGART 80, GERMANY

Introduction.

The Chemistry of diorganyl zinc derivatives is well investigated, mainly due to the easy access of these compounds and the variety of applications in organic, metalorganic and inorganic chemistry as mild alkylating reagents[1], in the stereospecific and stereoselective addition to imines and ketones[2] as well as for the preparation of zincates.[3] The synthesis and the spectroscopic properties as well as the molecular structures of zincates are only sparingly investigated despite their wide preparative use. Weiss and coworker solved the structures of powders of solvent-free Dilithium-tetramethylzincate $Li_2[ZnMe_4]$[4] and Dipotassium-tetraethinylzincate $K_2[Zn(C{\equiv}CH)_4]$[5] but in regard to rather high standard deviations the discussion of the Zn-C bond lengths is limited. The steric demand of the organyl substituent and the stoichiometry influences whether a tri- or a tetraorganylzincate is isolated. Thus $K[ZnMe_3]$ with a trigonal planar coordinated zinc atom shows Zn-C bond distances of about 207 pm.[6]

Results.

Synthesis. Bis[bis(trimethylsilyl)methyl]zinc (1) can easily be prepared via the metathesis reaction of Bis(trimethylsilyl)methyl-magnesiumchloride·2thf or Lithium-bis(trimethylsilyl)methanide with zinc(II)-chloride in diethylether. After removal of precipitated lithium chloride, the pyrophoric compound 1 with a melting point of only -89°C[7] is distilled at 70-72°C/4·10^{-2} Torr. During this synthesis no formation of a zincate has been established.

Alternatively Dilithium-tetramethylzincate[3,4] can be isolated after addition of dimethylzinc to an etheral solution of lithium-methanide and subsequent removal of all volatile material at 50°C/10^{-3} Torr. For the preparation of lithium tris[bis(trimethylsilyl)methyl]zincate a chelating complex ligand has to be added to the reaction solution. In ether solution the 1/1 mixture of lithium bis(trimethylsilyl)methanide and Bis[bis(trimethylsilyl)methyl]zinc (1) is treated at room temperature with one equivalent of tmeda (bis(dimethylamino)ethane); the resulting Lithium-tris[bis(trimethylsilyl)methyl]zincate·tmeda·2Et$_2$O (**2a**) is recrystallized from diethylether. The addition of another tmeda leads to the formation of the ether-free zincate **2b** (Eq. (1)), whereas the tridentate tmta (1,3,5-trimethyl-1,3,5-triazinane) results in thf solution in the formation of Lithium-tris[bis(trimethylsilyl)methyl]zincate·tmta·thf (**2c**). All of these zincates have a four-coordinated lithium cation in common.

$$LiR + ZnR_2 \xrightarrow{+\ tmeda} Li[ZnR_3]\cdot tmeda\cdot 2Et_2O \xrightarrow{+\ tmeda} Li[ZnR_3]\cdot 2tmeda \quad (1)$$

$$\underset{\mathbf{1}}{} \qquad\qquad\qquad \underset{\mathbf{2a}}{} \qquad\qquad\qquad \underset{\mathbf{2b}}{}$$

$$R = CH(SiMe_3)_2$$

A convenient one-step synthesis starting with zinc(II)chloride and a threefold stoichiometric amount of lithium bis(trimethylsilyl)methanide yields also quantitatively the zincate in the presence of a chelating ligand such as tmeda or tmta without the necessity of isolating the diorganylzinc **1**.

The amino ligands tmeda and tmta coordinate at the Bis[bis(trimethylsilyl)methyl]zinc (**1**) in an unidentate manner, whereas the 2,2'-bipyridyl substituent appears to be a bidentate ligand.[8] In contrast to these complexes ZnR$_2$·L with L as tmeda (**3a**) or tmta (**3b**) the sterically less crowded dimethylzinc coordinates to two tmta molecules, both bonded in the unidentate manner, to yield ZnMe$_2$·2tmta with a melting point of 37°C.[9] In contrast the 2,2'-Bipyridyl-bis[bis(trimethylsilyl)methyl]zinc (**3c**) displays a four-coordinated metal center.[8] All of these complexes ZnR$_2$·L (**3a-c**) are soluble in benzene, only the derivatives **3a** and **3b** are dissociated. The equilibrium according to Eq. (2) in thf is already well-known for the 2,2'-bipyridyl adduct **3c**. The zincates **2a-2c** precipitate in hydrocarbons and are only soluble in ethers like thf or 1,2-dimethoxyethane.

$$\underset{\mathbf{3c}}{R_2Zn\cdot bpy} \xrightleftharpoons{<\ thf\ >} \underset{\mathbf{1}}{ZnR_2} + bpy \quad (2)$$

The presence of a Lewis base is inevitable for the genesis of a tris[bis(trimethylsilyl)methyl]zincate; already Isobe and coworker[10] published the synthesis of derivatives Li[ZnR$_3$] with R as methyl, butyl, phenyl and 1-butynyl in thf starting with the zinc(II)chloride·tmeda.

With an addition of the aromatic chelating Lewis base bpy (2,2'-bipyridine) to the reaction solution according to Eq. (1), a deep red solution evoked by a radical is observed. The same 2,2'-Bipyridyl-bis(trimethylsilyl)methylzinc radical **4** can be photolytically generated (Eq. (3)), however, the radicals {R·} have not been detected.

$$\underset{\mathbf{3c}}{R_2Zn\cdot bpy} \xrightarrow{h\nu} \underset{\mathbf{4}}{[RZn\cdot bpy]^{\cdot}} + \{R^{\cdot}\} \quad (3)$$

Also the reduction of derivative **3c** with potassium in thf yields the same radical **4** in agreement with the reaction mechanism (Eq. (4)) Kaim[11] proposed.

$$R_2Zn\cdot bpy + ZnR_2 \rightleftharpoons [bpy\cdot ZnR]^+ [ZnR_3]^- \xrightarrow{+\ K} [bpy\cdot ZnR]^{\textstyle\cdot} + K[ZnR_3] \quad (\ 4\ )$$

$$\text{3c} \qquad \text{1} \qquad\qquad\qquad\qquad\qquad\qquad\qquad \text{4}$$

The Bis[bis(trimethylsilyl)methyl]zinc (1) originates from the equilibrium displayed in Eq. (2). The charge transfer in form of the carbanion R^- from $R_2Zn\cdot bpy$ **3c** to compound **1** yields the zincate anion and a 2,2'-bipyridyl-alkylzinc cation; the latter is reduced by potassium to give radical **4**. Instead of potassium the lithium-bis(trimethylsilyl)methanide acts in our case as the reducing reagent (Eq. (5)).

$$[bpy\cdot ZnR]^+ [ZnR_3]^- + LiR \longrightarrow [bpy\cdot ZnR]^{\textstyle\cdot} + Li[ZnR_3] + \{R^{\textstyle\cdot}\} \qquad (\ 5\ )$$

$$\text{4}$$

NMR Spectroscopy. Table 1 displays the NMR parameters of the bis[bis(trimethylsilyl)methyl]zinc moiety R_2Zn of the compounds described above. For comparison reasons, the alkylating reagents Lithium-bis(trimethylsilyl)methanide and Bis(trimethylsilyl)methyl-magnesium-chloride·2thf are included. The α-CH group connected to the metal center responds sensitively to changes of the metal atom and the coordination sphere of the zinc atom. The chemical shifts $\delta(^1H)$ of the compounds $LiCH(SiMe_3)_2$ and $(Me_3Si)_2CH\text{-}MgCl\cdot 2thf$ are observed at about -2 ppm, whereas the corresponding value for a $(Me_3Si)_2CH$-ligand bonded at a zinc atom is found at about -0.7 ppm.

The α-carbon atoms connected to the metal center usually are registered at rather high field, but the substitution of the hydrogen atoms against trimethylsilyl groups leads to an extraordinary low field shift of the appropriate carbon atom [12] combined with a decreasing $^1J(CH)$ coupling constant. The solvate-free Bis[bis(trimethylsilyl)methyl]zinc (1) [12] and its tmeda and tmta adducts exhibit $\delta(^{13}C)$-values around 14 ppm, whereas the corresponding α-C-atom of the 2,2'-bipyridyl complex **3d** has a remarkable high field shift of 5 ppm. This exception can be explained by the greater coordination number of 4 for the zinc atom in this derivative; on the other hand 2,2'-bipyridine is a stronger base than the trialkylamine ligands such as tmeda or tmta. The lithium tris[bis(trimethylsilyl)methyl]zincates **2a**, **2b** and

Table 1 NMR data for the α-CHSi$_2$-moiety bonded to a metal center (R = CH(SiMe$_3$)$_2$)

	Solv.	$\delta(^1H)$	$\delta(^{13}C)$	$\delta(^{29}Si)$	$^1J(SiC)$	$^1J(CH)$	Ref.
RMgCl	thf-d$_8$	-1,90	-1,89	-3,39	43,8	96,8	
RLi	C$_6$D$_6$	-2,14	1,44	-6,36	41,2	94,3	
1	C$_6$D$_6$	-0,60	13,97	-2,50	41,9	102,8	7
3a	C$_6$D$_6$	-0,62	13,89	-2,53	41,9	104,2	
3b	C$_6$D$_6$	-0,59	13,87	-2,53	41,8	104,1	
3c	C$_6$D$_6$	-0,75	4,8	-1,88	46,6	117,4	8
2b	thf-d$_8$	-0,86	9,20	-2,80	43,8	102,5	
2c	thf-d$_8$	-1,06	7,18	-3,50	45,9	97,4	

Table 2 Selected structural parameters of the compounds $R_2Zn\cdot bpy$ **3c**, $R_2Zn\cdot tmta$ **3b** and $Li[ZnR_3]\cdot tmeda\cdot 2Et_2O$ **2a** ($R = CH(\check{S}iMe_3)_2$).

	ZnC	ZnN	$C_\alpha Si$	$C_{Me}Si$	CZnC	CZnN	Ref.
3c	203	219	183	187	126	107–116	8
3b	199	239	186	187	157	100/102	
2a	209	—	184	188	120	—	

2c show $\delta(^{13}C)$-values of about 8 ppm

The $^{29}Si\{^1H\}$ chemical shifts of $ZnR_2\cdot L$ **3** and $Li[ZnR_3]\cdot 2L$ **2** appear to be grouped around a value of -3 ppm whereas the smaller values originate from the zincates.

Structures. In order to understand why the chelating tertiary aliphatic amines support the formation of the zincates and the aromatic amine leads to the generation of radicals we have to compare some structural parameters (Table 2). It is evident that the bypyridine is bonded much stronger to the zinc atom than the unidentate tmta ligand; the Zn–N-bond lengths differ of 20 pm. The slightly elongated Zn–C-bond of derivative **3c** is a consequence of the higher coordination number 4. The loose coordination of tmta causes a large CZnC-angle of 157°, so that in a slightly exaggerated point of view the zinc atom can be regarded as sp-hybridized with a tmta molecule bonding into an empty p-orbital.

The zincate anion shows a trigonal planar coordinated zinc atom. The long Zn–C-bonds of 209 pm can be explained by the steric demand of the bis(trimethylsilyl)methyl substituents, reinforced by an electrostatic repulsion of the three partially negatively charged alkyl ligands. The zincate anion nearly displays C_3-symmetry with a propeller-like orientation of the $(Me_3Si)_2CH$-groups. The coordination sphere of the α-carbon atoms is flattened as already indicated by the small $^1J(^{29}Si-^{13}C)$ and $^1J(^{13}C-^1H)$ coupling constants.

References.
1. S. Moorhouse, G. Wilkinson; J. Organomet. Chem., 1973, 52, C5.
2. (a) W. Tückmantel, K. Oshima, H. Nozaki; Chem. Ber., 1986, 119, 1581.
   (b) R. Noyori, M. Kitamura; Angew. Chem., 1991, 103, 34.
3. J. Yamamoto, C. A. Wilkie; Inorg. Chem., 1971, 10, 1129.
4. E. Weiss, R. Wolfrum; Chem. Ber., 1968, 101, 35.
5. E. Weiss, H. Plass; J. Organomet. Chem., 1968, 14, 21.
6. P. v. Ragué Schleyer, C. Schade; Adv. Organomet. Chem., 1987, 27, 169.
7. M. Westerhausen, B. Rademacher, W. Poll; J. Organomet. Chem., 1991, 421, 175.
8. M. Westerhausen, B. Rademacher, W. Schwarz; J. Organomet. Chem., 1992, 427, 275.
9. M. B. Hursthouse, M. Motevalli, P. O'Brien, J. R. Walsh; Organometallics, 1991, 10, 3196.
10. M. Isobe, S. Kondo, N. Nagasawa, T. Goto; Chem. Lett., 1977, 679.
11. W. Kaim; Chem. Ber., 1981, 114, 3789.

Low Valent Coinage Metal Coordination Compounds with Tertiary Phosphines and Thiones

P. D. Akrivos, P. Karagiannidis, P. Aslanidis,
G. S. Kapsomenos, and S. K. Hadjikakou

ARISTOTLE UNIVERSITY OF THESSALONIKI, FACULTY OF
CHEMISTRY, GENERAL AND INORGANIC CHEMISTRY DEPARTMENT,
PO BOX 135, GR-540 06 THESSALONIKI, GREECE

1 INTRODUCTION

The utility of copper(I) compounds in approximating the active site of several copper-containing proteins has prompted us to investigate the structure and reactivity of a series of mixed ligand copper(I) compounds with heterocyclic thiones, possessing at least one α-nitrogen heteroatom and tertiary phosphines. In this respect, both aromatic (e.g. pyridine-2-thione, pyrimidine-2-thione, thiazolidine-2-thione, imidazolidine-2-thione) and saturated (tetrahydropyrimidine-2-thione and ω-thiocaprolactam) thiones have been used, the latter introducing a substantial bulk to the coordination environment of the metal. The phosphines used so far involve triphenylphosphine, the tritolyl phosphines, which offer a wide range of bulk (especially the tri-o-tolylphosphine with a cone angle of 177°) as well as the very bulky but saturated tricyclohexyl phosphine.

The compounds studied form two major categories, with respect to their stoichiometry; the $[Cu(PR_3)(thione)X]_m$ and the $Cu(PR_3)_n(thione)_{3-n}X$ where X= Cl, Br or I and n= 1 or 2. In the first category, both monomer (m=1) and dimer (m=2) compounds are incorporated, while the second one consists of monomer compounds. The crystal structure determination of these complexes is essential for their study; nevertheless, conventional spectroscopic and other techniques have been useful in their characterization and these findings will be presently discussed.

2 PREPARATION OF THE COPPER(I) COMPOUNDS

The usual way of preparation of the compounds studied is by means of direct addition of the appropriate thione and phosphine ligands in the desirable molar ratio to acetonitrile solutions of the copper(I) halides. In this way, the reaction is facilitated, since the already existing $Cu(CH_3CN)_nX$ compounds readily exchange the CH_3CN ligands with thione and phosphine molecules. A variation was unavoidable in the case of the saturated thiones, where it was observed that their addition should precede that of the phosphine ligands in order to obtain mixed ligand complexes. The precursor compounds may therefore be classified as:

$Cu(thione)_nX$ for saturated thiones and
$Cu_mX_m(PR_3)_n$ for aromatic thiones

3 PRECURSOR COPPER(I) COMPOUNDS

The characteristics of these precursor compounds are interesting and will be discussed in relation to the structures of the final products.

The $Cu(thione)_nX$ compounds are three-coordinate species[1]. The bromide complexes appear to be more pyramidal than their chloride and iodide counterparts, a fact endorsed also by semiempirical molecular orbital calculations[2] on model compounds, where standard Cu-X, Cu-P and Cu-S bond lengths were taken into consideration. All these complexes are reactive towards both Lewis bases and small labile molecules (e.g. CO, CO_2, CS_2). A CO_2 insertion product was produced[1b] by room temperature addition of CO_2 at 1 atm. pressure to $Cu(thiocaprolactam)_2Br$.

Up to now, a single structure of a phosphine precursor has been determined[3], namely that of $Cu_2I_2\{P(m\text{-tolyl})_3\}_3$, where the two copper atoms adopt different local environments. Semiempirical computations on model compounds of this series[3] show that $Cu_2X_2(PR_3)_3$ should be more stable than $[CuX\{PR_3\}_2]_2$ and almost as reactive as $[CuX\{PR_3\}]_2$; therefore being more suitable as precursor unless extreme steric effects are present.

4 MIXED LIGAND COPPER(I) COMPLEXES

Structural considerations

The products obtained from $Cu(thione)_nX$ precursors are monomer compounds, either three- or four- coordinate depending on the steric effects of the ligands. In the case of the bulky thiocaprolactam, triphenylphosphine gave rise to tetrahedral compounds[4] [$Cu(thione)_2(PPh_3)X$ or $Cu(thione)(PPh_3)_2X$] depending on the PPh_3-to-thione molar ratio. When tricyclohexylphosphine was used, monomer compounds of the stoichiometry $Cu(thione)(Pcy_3)X$ were obtained[2]. The tendency of bromine complexes to adopt "pyramidal" conformation is once more supported by the structural studies so far carried out as well as theoretical considerations.

The compounds obtained by $Cu_mX_m(PR_3)_n$ intermediates are generally dimeric in nature except for those involving tri-o-tolyl phosphine (totp), which, due to the steric effect of the phosphine, are monomers of the formula $Cu(thione)(totp)X$[5]. In all other cases a Cu_2L_2 core is formed and the environment around both copper atoms is pseudotetrahedral. The bridging atoms L are either halogen or sulfur atoms, depending on the "hardness" of the attached thione and halogen atoms. In all chlorine compounds, regardless of the phosphine and thione present, the core is Cu_2S_2, while in the iodo compounds studied the core is Cu_2I_2[6]. Bromine, representing a borderline case, may engender either of the above cores; nevertheless, semiempirical computations predict[7] that formation of Cu_2Br_2 core should be preferable due to orbital stabilization relative to Cu_2S_2. This is, though, a point that needs advocation by further experimental work and more elaborate calculations. A few interesting points concerning the spectral characterization of the copper(I) compounds as well as their reactivity are presented in the following section.

Spectral and reactivity studies

The ^{1}H n.m.r. focusing mainly on the distinct NH- hydrogen signal of the coordinated thiones as well as ^{31}P n.m.r. recorded in CDCl$_3$ solutions reveal small shifts to higher fields on going from chloro to iodo complexes, pursuant to the shielding effect of the halogen atoms. The ^{31}P spectra show single lines and only at -70° C a splitting was observed giving rise to a poorly resolved signal originating from the interaction of P and Cu nuclei[8].

The photoreactivity of the compounds was studied by irradiating chloroform solutions at 304 nm with a high pressure Hg lamp and recording the spectra at set time intervals. Coleman's method was applied and revealed the presence of two photoproducts in the solution. These products were identified as PR$_3$ and the [Cu(thione)X]$_2$ complex in the case of the dimer compounds or PR$_3$ and [Cu(thione)X]$_n$ in the case of monomer complexes. The irradiation, therefore, must have promoted the reaction

$$[Cu(PR_3)(thione)X]_n \rightleftharpoons [Cu(thione)X]_n + n\,PR_3$$

Since the λ_{exit} used corresponds roughly to the $\pi \rightarrow \pi^*$ excitation of the thione ligands it is natural that dimer compounds with Cu$_2$X$_2$ cores give higher quantum yields than those with Cu$_2$S$_2$ ones, while in the case of monomer compounds the major factor is the relative strength of the Cu-S and Cu-P bonds.

Finally, electrochemical reduction of the complexes in acetonitrile solutions with NBu$_4$BF$_4$ as supporting electrolyte at scan rates between 0.2 and 1 V/sec showed three irreversible reduction peaks at *ca.* -1.0, -1.7 and -2.1 V *vs.* SCE for the monomer and four (*ca.* -1.0, -1.6, -1.8 and -2.1 V) for their dimer counterparts[9]. The last one in every case corresponds to the reduction of the thione ligand. In every case, copper deposition was evident at the surface of the glassy carbon electrode used. Besides the well known dissociation of dimer compounds to their monomer "fragments" the exchange reaction between phosphine ligand and solvent molecules, described below, may be responsible for the presence of various species in the solution.

$$[Cu(PR_3)(thione)X]_n + n\,CH_3CN \rightleftharpoons [Cu(thione)(CH_3CN)X]n + nPR_3$$

5 SILVER(I) AND GOLD(I) COMPLEXES

The silver(I) compounds were prepared by the phosphine intermediate route[10] and their spectral characteristics account for monomer compounds with tetrahedral environments, their stoichiometry being determined by the phosphine: thione molar ratio. Bulky phosphines have not yet been used but are expected to produce three-coordinate species, whose reactivity must be interesting. All the Ag(I) complexes prepared are moderately light sensitive.

Gold(I) complexes were prepared by direct reduction of Au(III) with a four-fold excess of the appropriate thione ligand in THF or methanol. The only structurally characterized Au(I) compound is that of thiazolidine-2-thione[11], which may be classified among epitomes of linear Au(I) coordination environment. The Au(thione)$_2^+$ units are involved in an extended hydrogen bond network, in which both the chloride counterion and the water molecule present in the unit cell participate to a significant extent. The preference for *cis* orientation of the thione ligands as well as the unique torsion angle of 40˚ between the two thione molecules in the present case are being investigated computationally and further experimentation is being done with the bulkier saturated thiones.

6 CONCLUDING REMARKS

The bulk of the thione and phosphine ligands is important in determining the formation of the monomer or dimer Cu(I) compounds, while their "hardness" is crucial for the kind of dimer formed. The structures adopted by the precursor compounds, whether determined crystallographically or anticipated computationally, may provide valuable information for the ultimate preparation of mixed ligand compounds. Spectroscopic and other techniques are not all adscititious, in the sense that, carefully and thoughtfully applied and in connection with appropriate computation, they may afford a better understanding of the nature and reactivity of monomer and oligomer coordination compounds of the coinage metals; such knowledge is a necessary prerequisite for the approximation of reactions in vivo, where these metals play an indispensable role.

REFERENCES

1. a) P. Karagiannidis, P.D. Akrivos, A. Hountas, A. Terzis, Inorg. Chim. Acta, 1990, 180, 93; b) P. Karragiannidis, P.D. Akrivos, B. Koijc–Prodic, M. Luic, J. Coord. Chem., accepted for publication.
2. P.D. Akrivos, G. Kapsomenas, P. Karagiannidis, A. Aubry, S. Skoulika, Inorg. Chim. Acta, 1992, 202, 73.
3. P.D. Akrivos, S.K. Hadjikakou, P. Karagiannidis, D. Mentzafos, A. Terzis, Inorg. Chim. Acta, submitted.
4. B. Koijc–Prodic, M. Luic, P. Karagiannidis, P.D. Akrivos. S. Stoyanov, J. Coord. Chem., 1992, 25A, 21.
5. S.K. Hadjikakou, P. Aslanidis, P. Karagiannidis, A. Aubry, S. Skoulika, Inorg. Chim Acta, 1992, 193, 129.
6. a) S.K. Hadjikakou, P. Aslanidis, P. Karagiannidis, A. Hountas, A. Terzis, Inorg. Chim. Acta, 1991, 184, 161; b) S.K. Hadjikakou, P. Aslanidis, P. Karagiannidis, D. Mentzafos, A. Terzis, Polyhedron, 1991, 9, 935.
7. S.K. Hadjikakou, P. Aslanidis, P.D. Akrivos, P. Karagiannidis, B. Koijc–Prodic, M. Luic, Inorg. Chim. Acta, accepted for publication.
8. P. Karagiannidis, P.D. Akrivos, J. Herema, M. Luic, B. Koijc–Prodic, work in progress.
9. P. Karagiannidis, S.K. Hadjikakou, N. Papadopoulos, Polyhedron, accepted for publication.
10. P. Karagiannidis, P. Aslanidis, S. Kokkou, C.J. Cheer, Inorg. Chim. Acta, 1990, 172, 247.
11. P. Karagiannidis, P.D. Akrivos, M. Gdaniec, Z. Kosturkiewicz, Inorg. Chim. Acta, submitted.

An Unusual Copper(II) Complex Containing Both Halide and Pseudohalide Coordinated Ligands: [Cu(terpy)Cl(N₃)]

R. Cortés, L. Lezama, J. I. R. Larramendi, J. L. Mesa, G. Madariaga[1], and T. Rojo

DEPARTAMENTOS DE QUÍMICA INORGÁNICA Y FÍSICA DE LA MATERIA CONDENSADA[1], UNIVERSIDAD DEL PAÍS VASCO, APTDO. 644, 48080 BILBAO, SPAIN

INTRODUCTION

The halide and pseudohalide ions can coordinate to metals as monodentate (terminal) or bidentate (bridging) ligands. Pseudohalide anions readily displace halide ions from their compounds. This characteristic is usually employed in different synthetic strategies and means that the coexistence of both different anions in the same compound is most unusual. As far as we are aware, this is the first Cu(II) complex containing both halide (Cl) and pseudohalide (N₃) ligands.

EXPERIMENTAL

Synthesis

The title compound has been synthesized by using as starting product the [Cu(terpy)Cl₂] (terpy = 2,2':6',2"–terpyridine) complex,[1] and as a source of azide anions the [Cu(terpy)(N₃)₂] complex.[2] Stoichiometric amounts were employed and good quality crystals were obtained by slow evaporation, at room temperature, of the aqueous solution formed. *Anal. Calcd* for $C_{15}H_{11}N_6CuCl$: C, 48.1; H, 3.0; N, 22.5; Cu, 17.0. *Found*: C, 48.0; H, 2.9; N, 22.7; Cu, 16.9.

X–ray Structure Determination

Crystals of [Cu(terpy)Cl(N₃)] as green prisms were mounted on an Enraf–Nonius CAD4 automatic diffractometer. The cell dimensions and the orientation matrix were determined from the setting angles of 25 reflections with Mo Kα monochromated radiation. A fast data collection of reflections in the θ range 3–12° allowed us to choose the space group on looking at the systematic absences. Monoclinic system, space group $P2_1/n$, with a = 10.586(2), b = 8.572(1), c = 16.396(1)Å, β = 100.69(5)°, Z = 4. Orientation and intensity control by means of two standard reflections showed no significant changes during data collection. The structure was solved using direct methods (MULTAN 84)[3] and successive Fourier synthesis (SHELX76).[4] Anisotropic thermal parameters were introduced for all non–hydrogen atoms. Final values of the discrepancy indices are R = 0.036 (R_w = 0.036) for 4627 independent reflections with I$\geqslant$2.5σ(I).

RESULTS AND DISCUSSION

Description of the structure

The structure consists of discrete molecules in which the coordination polyhedron around the copper(II) ion can be described as a distorted square pyramid. The three nitrogen atoms of the terpy ligand (N1, N2, N3) and one nitrogen of the azide group (N4) occupy the basal positions, with the chloride atom in the apical one (see Figure 1).

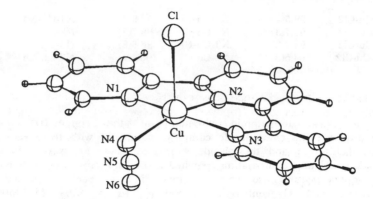

Figure 1. Molecular structure of the [Cu(terpy)Cl(N$_3$)] compound

Table 1. Fractional atomic coordinates (x10^4, x10^5 for Cu,Cl) for the title compound

ATOM	X	Y	Z
Cu	70070(3)	7732(3)	8485(2)
Cl	84786(5)	29897(7)	12316(3)
N1	8386(2)	-958(2)	1033(1)
C11	8905(2)	-1667(3)	1740(1)
C12	9802(3)	-2849(3)	1784(1)
C13	185(2)	-3318(3)	1067(2)
C14	9685(2)	-2571(3)	328(1)
C15	8787(2)	-1407(2)	331(1)
N2	7257(2)	428(2)	-292(1)
C21	8170(2)	-564(2)	-424(1)
C22	8455(2)	-748(3)	-1214(1)
C23	7768(3)	148(3)	-1848(1)
C24	6810(2)	1137(3)	-1709(1)
C25	6562(2)	1252(2)	-910(1)
N3	5506(2)	2038(2)	184(1)
C31	5542(2)	2180(2)	-640(1)
C32	4682(2)	3122(3)	-1154(2)
C33	3753(2)	3912(3)	-824(2)
C34	3705(2)	3753(3)	7(2)
C35	4601(2)	2800(3)	492(2)
N4	6208(2)	625(3)	1835(1)
N5	6550(2)	7(3)	2424(1)
N6	6851(3)	-639(4)	3062(2)

Table 2 Selected bond distances (Å) and angles (°) for the title and related compounds

[Cu(terpy)Cl$_2$]		[Cu(terpy)Cl(N$_3$)]		[Cu(terpy)(N$_3$)$_2$]	
Cu-N1	2.056(5)	Cu-N1	2.064(7)	Cu-N1	2.05(1)
Cu-N2	1.952(5)	Cu-N2	1.959(8)	Cu-N2	1.94(1)
Cu-N3	2.052(5)	Cu-N3	2.060(5)	Cu-N3	2.06(1)
Cu-Cl2	2.469(2)	Cu-Cl	2.463(4)	Cu-N4	2.21(2)
Cu-Cl1	2.252(2)	Cu-N4	1.964(5)	Cu-N7	1.96(2)
N1-Cu-Cl2	99.5(4)	N1-Cu-Cl	96.9(4)	N1-Cu-N4	98.3(5)
N2-Cu-Cl2	98.7(3)	N2-Cu-Cl	99.7(3)	N2-Cu-N4	97.0(7)
N3-Cu-Cl2	92.4(3)	N3-Cu-Cl	96.8(3)	N3-Cu-N4	94.3(5)
Cl1-Cu-Cl2	104.5(3)	N4-Cu-Cl	100.9(4)	N7-Cu-N4	99.1(7)

The main interatomic distances of the copper(II) coordination polyhedron for the chloro–azide complex, compared with those corresponding to the dichloro [1] and the diazide [2] related ones, are shown in Table 2. Distortion of the coordination polyhedra of the three related compounds from square pyramid to trigonal bipyramid has been calculated by the Muetterties and Guggenberger method [5] [Δ = 0.68 (dichloro), 0.81 (chloro–azide), 0.87 (diazide)]. It can be observed, in all cases, that the topologies are close to square pyramidal, the most distorted one corresponding to the dichloro compound.

EPR Spectroscopy

The X–band powder EPR spectra for the three compounds are shown together in Figure 2. Though all three mean signals have an orthorhombic symmetry for the g tensor, clear differences can be observed between them. This orthorhombic symmetry is not in contradiction to the square pyramidal topology for the copper coordination polyhedra, because of the existence of different bond distances between the atoms involved.

The coordination sphere of the [Cu(terpy)Cl$_2$] compound shows three short ($\sim$2 Å), one medium (2.25 Å) and one large (2.47 Å) distance. Its EPR spectrum has a clear orthorhombic symmetry with g_1 = 2.18, g_2 = 2.13, g_3 = 2.07.

The [Cu(terpy)Cl(N$_3$)] and [Cu(terpy)(N$_3$)$_2$] coordination spheres show four short (slightly different two by two: 2.05, 1.95 Å) and one large distance, obviously greater in the chloro–azide compound (2.46 Å) than in the diazide one (2.21 Å). Their EPR spectra also have an orthorhombic symmetry for the g tensor, being less defined than for the dichloro complex, owing to the minor difference between the bond distances.

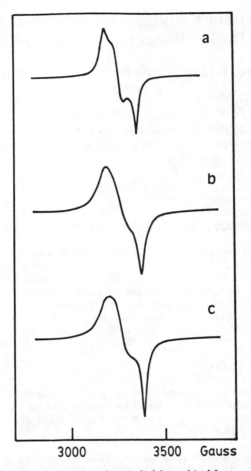

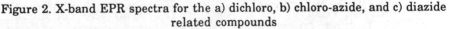

Figure 2. X-band EPR spectra for the a) dichloro, b) chloro-azide, and c) diazide related compounds

The corresponding "g" values for the chloro–azide compound are g_1 = 2.19, g_2 = 2.12, g_3 = 2.06, while for the diazide one g_1 = 2.18, g_2 = 2.12, g_3 = 2.05. No significant differences between the g values are observed for all the compounds studied.

REFERENCES

1. W. Henke, S. Kremer and D. Reinen, *Inorg. Chem.*, 1983, <u>22</u>, 2858.
2. J. Via, M.I. Arriortua, T. Rojo, J.L. Mesa and A. Garcia, *Bull. Soc. Chim. Belg.*, 1989, <u>98</u>, 179.
3. P. Main, G. Germain and M.M. Woolfson, MULTAN 84, a system of computer programs for the automatic solution of crystal structures from X–ray diffraction data, Universities of York and Louvain, 1984.
4. G.M. Sheldrick, SHELX 76, Program for Crystal Structure Determination, University of Cambridge, England 1976.
5. E.L. Muetterties and L.J. Guggenberger, *J. Amer. Chem. Soc.*, 1974, <u>96</u>, 1748.

Development of Homogeneous Electrocatalysts Based on Copper for the Reduction of Carbon Dioxide

J. S. Field[1], R. J. Haines[1], C. P. Kubiak[2], C. Parry[1], S. Reiser[1], S. Sookraj[1], and R. E. Wittrig[2]

1 DEPARTMENT OF CHEMISTRY, UNIVERSITY OF NATAL, PO BOX 375, PIETERMARITZBURG 3200, SOUTH AFRICA

2 DEPARTMENT OF CHEMISTRY, PURDUE UNIVERSITY, WEST LAFAYETTE, IN 47906, USA

1 INTRODUCTION

Carbon dioxide is a potential chemical feedstock but its conversion to useful organic products requires a process that involves at least one reduction step. The reduction can be effected electrochemically but in practice any such reduction is normally accompanied by a large overvoltage. Electrocatalysts can be employed to reduce this overvoltage and a wide range of homogeneous systems have been investigated in this context.[1,2] Mononuclear metal bipyridyl compounds have been the most extensively studied,[2] particularly mechanistically, and a number have proved to be very effective, $[RuH(CO)(bipy)_2]PF_6$ being a typical example.[3] However, in all reported cases involving systems of this type, the reduction of CO_2 has not been effected beyond the two-electron stage.

It was anticipated that dinuclear and metal cluster bipyridyl derivatives might be more effective as electron reservoirs thereby catalyzing the reduction of CO_2 beyond CO or $HCOO^-$ while, furthermore, it was considered that the co-ordination of CO_2 across two adjacent metal sites as previously established for $[Ru_2\{\mu-\eta^2-OC(O)\}(CO)_4\{\mu-(Pr^iO)_2PN(Et)P(OPr^i)_2\}_2]^4$ could well lead to its increased activation. The phosphorus-bipyridyl ligand

has thus been synthesized with the object of producing a ligand containing a chelating bipyridyl fragment but which overall will preferentially adopt a bridging co-ordination mode thereby stabilizing dinuclear compounds to fragmentation.

2 RESULTS AND DISCUSSION

Treatment of $[Cu(MeCN)_4]PF_6$[5] in acetonitrile with an equimolar amount of this phosphorusbipyridyl ligand was found to lead to the rapid formation of the dicopper complex $[Cu_2(\mu-PPh_2bipy)_2(MeCN)_2](PF_6)_2$ (1), characterized X-ray crystallographically (Fig. 1) as well as by conventional methods. The acetonitrile ligands in this complex are labile and readily replaced by other neutral and anionic ligands. Thus benzonitrile, pyridine and water for instance readily substitute the acetonitriles to afford $[Cu_2(\mu-PPh_2bipy)_2(PhCN)_2]-(PF_6)_2$, $[Cu_2(\mu-PPh_2bipy)_2(py)_2](PF_6)_2$ and $[Cu_2(\mu-PPh_2bipy)_2(H_2O)_2](PF_6)_2$ respectively, all of which have been characterized crystallographically. Significantly, while attack by cyanide ions leads to the formation of the neutral species $[Cu_2(CN)_2(\mu-PPh_2bipy)_2]$, halide and carboxylate ion attack produces the halo- and carboxylato-bridged species, $[Cu_2(\mu-X)(\mu-PPh_2bipy)_2]PF_6$ (X = Cl, Br or I) and $[Cu_2\{\mu-\eta^2-OC(R)O\}(\mu-PPh_2bipy)_2]PF_6$ (R = H, Me, etc.) respectively. The phosphorusbipyridyl ligand has also been reacted with the copper bipyridyl precursor $[Cu(bipy)(MeCN)_2]PF_6$ and found to afford the novel asymmetric dicopper complex $[Cu_2(\mu-PPh_2bipy)_2(bipy)](PF_6)_2$ (Figure 2).

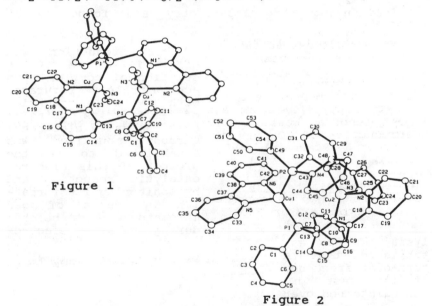

Figure 1

Figure 2

The cyclic voltammogram of $[Cu_2(\mu-PPh_2bipy)_2(MeCN)_2](PF_6)_2$ has been measured under argon and found to contain two reduction peaks at $E_{\frac{1}{2}}(2+/+) = -1.34V$ vs Ag/AgCl and $E_{\frac{1}{2}}(+/o) = -1.55V$ vs Ag/AgCl. The one-electron nature of both reductions was confirmed by chronoamperometry. The first reduction is fully reversible with a peak to peak separation, ΔE, of 65mV while the second reduction also exhibits reversibility with ΔE = 80mV (Figure 3). A similar voltammogram was obtained for $[Cu_2(\mu-PPh_2bipy)_2(py)_2]PF_6$ (2). Significantly, while the primary reduction peak remained essentially unperturbed when the cyclic voltammograms of these two compounds were measured under an atmosphere of carbon dioxide, the second reduction peak revealed marked current enhancement under these conditions. Furthermore, the oxidation peaks in the reverse scan were no longer present. These observations are indicative of these compounds functioning as electrocatalysts for CO_2 reduction. The original cyclic voltammograms were again obtained on purging the solution with argon for several minutes.

The heterogeneous electron transfer rates for the two reduction steps involving compound **2** in acetonitrile were investigated employing rotating disc voltammetry (Figure 4). Using a platinum rotating electrode, a Levich plot was generated for each reduction. From the intercept for the first reduction it was possible to calculate a K_e value of 0.04 cm s^{-1}; the second reduction proved too rapid to obtain a rate constant by this technique. The homogeneous rate constant, k_{CO_2}, for the reduction of CO_2 by complex **1** was obtained using chronoamperometry and found to be 1.1 $M^{-1}s^{-1}$.

Cyclic Voltammogram of [Cu$_2$(μ-Ph$_2$Pbipy)$_2$(MeCN)$_2$](PF$_6$)$_2$

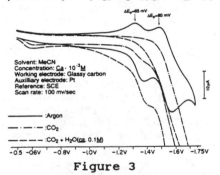

Figure 3

A bulk reduction was effected on a 0.3mM solution of complex **1** in acetonitrile saturated with CO_2 at a potential of -1.7 volts using a platinum gauze electrode. Carbon monoxide was established to be the product of this electrocatalytic reduction and by monitoring its formation by means of gas chromatography, before, during and after electro-

lysis, it was established that a turn-over frequency of at least 2h^{-1} was maintained over the course of 24 hours. The catalyst remained active after this period and by measuring an IR spectrum of the solid product following electrolysis it was established that the copper dimer was still present

Rotating Disc Voltammograms of [Cu$_2$(u-Ph$_2$Pbipy)$_2$(py)$_2$](PF$_6$)$_2$

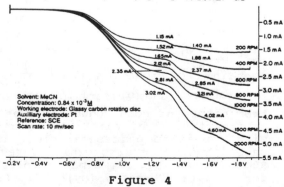

Figure 4

in its original form.

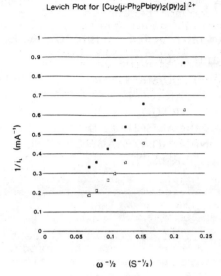

Levich Plot for $[Cu_2(\mu\text{-}Ph_2Pbipy)_2(py)_2]^{2+}$

$\omega^{-\frac{1}{2}}$ $(S^{-\frac{1}{2}})$

Figure 5

A catalytic cycle is proposed in which the first step is a two-stage reduction of the oxidized form of the catalyst. Displacement of an acetonitrile ligand by a CO_2 molecule is followed by insertion of a second CO_2 to afford a co-ordinated $OCOC(O)O$ group. Disproportionation of this group produces CO and CO_3^{2-} and leads to the regeneration of the oxidized form of the catalyst. Current studies include the development of methods for the polymerization of the catalyst onto electrode surfaces. This would not only lead to an increase of the local concentration of the catalyst, thereby increasing the rate of conversion to products, but would also create a potentially more durable catalytic system as well as possibly reducing any remaining overvoltage.

REFERENCES

1. J.R. Pugh, M.R.M. Bruce, B.P. Sullivan and T.J. Meyer, *Inorg. Chem.*, 1991, **30**, 86.
2. C.P. Kubiak and K.S. Ratliff, *Israel J. Chem.*, 1991, **31**, 3.
3. B.P. Sullivan, M.R.M. Bruce, T.R. O'Toole, C.M. Bolinger, E.Megehee, H. Thorp and T.J. Meyer, *Catalytic Activation of Carbon Dioxide*, Ed. W.M. Ayers, A.C.S. Symposium Series 363, American Chemical Society, Washington DC, 1988.
4. J.S. Field, R.J. Haines, J.Sundermeyer and S.F. Woollam, *J. Chem.Soc., Chem. Commun.*, 1990, 985.
5. G.J. Kubas, *Inorg. Syn.*, 1979, **19**, 90.

Charge-transfer Salts from Organometallic Gold(I) and Gold(III) Complexes

E. Cerrada[1], J. Garín[1], M. C. Gimeno[1], A. Laguna[1], M. Laguna[1,*], and P. G. Jones[2]

[1] DEPARTAMENTO DE QUÍMICA INORGÁNICA, INSTITUTO DE CIENCIA DE MATERIALES DE ARAGÓN, FACULTAD DE CIENCIAS, UNIVERSIDAD DE ZARAGOZA, CSIC 50009 ZARAGOZA, SPAIN

[2] INSTITUT FÜR ANORGANISCHE UND ANALYTISCHE CHEMIE DER TECHNISCHEN UNIVERSITÄT, HAGENRING 30, W-3300 BRAUNSCHWEIG, GERMANY

1. Introduction

There has been an increasing interest in the chemical and physical properties of the products of reaction between inorganic complexes and electron-withdrawing organic acceptors such as 7,7',8,8'-tetracyanoquinodimethane (TCNQ)[1] or organic donors such as tetrathiafulvalene derivatives (TTF's)[2,3]. In some cases, these materials exhibit higher than expected electrical conductivity, unique structures with unusual oxidation states, and unexpected magnetic properties.

In most of these compounds the non-organic part plays an important role in determining the stoichiometry and the arrangement of the TCNQ or TTF molecules in the solid state and, often, in governing a partial charge transfer to the organic moieties.

We have undertaken the systematic study of charge transfer complexes obtained by reaction of different gold(I) derivatives with LiTCNQ and TCNQ or by electrocrystallization procedures. The structures of [Au(PPh$_3$)$_2$TCNQ], [Au(CH$_2$PPh$_3$)$_2$(TCNQ)$_2$], [Au$_3$(CH$_2$PPh$_2$CH$_2$)$_2$(PPh$_3$)$_2$]TCNQ and [Ph$_2$TTF]$_2$[Au(C$_6$F$_5$)$_2$] have been established by X-ray diffraction.

2. Results and Discussion

In our first attempts we selected as precursors the complexes [Au(tht)L]ClO$_4$ (L= phosphine or ylide), which possess a labile ligand (tetrahydrothiophene) and reacted them with LiTCNQ. When L= phosphine the reaction leads to a mixture of products which can not be resolved by recrystallization. When L= ylide (i.e. L= CH$_2$PPh$_3$ or

CH_2PPh_2Me), the reaction in ethanol leads to a mixture of metallic gold and [Au(ylide)$_2$(TCNQ)$_2$]. The isolation of the gold complexes is easy owing to their solubility in dichloromethane.

Despite these results, this type of complex can not be obtained starting from [Au(CH$_2$PR$_3$)$_2$]ClO$_4$ by reaction with LiTCNQ, and in this case Au(CH$_2$PR$_3$)$_2$(TCNQ) are obtained instead of the TCNQ derivatives.

$$[Au(CH_2PR_3)_2]ClO_4 + LiTCNQ \longrightarrow Au(CH_2PR_3)_2(TCNQ)$$

R$_3$= Ph$_3$ **1a**, Ph$_2$Me **1b**, PhMe$_2$ **1c**

Complexes [Au(CH$_2$PR$_3$)$_2$(TCNQ)$_2$] can be obtained either by reacting complexes **1a-c** with TCNQ° or in a one pot synthesis by reacting the bisylide derivatives with LiTCNQ and TCNQ.

$$Au(CH_2PR_3)_2(TCNQ) + LiTCNQ \longrightarrow Au(CH_2PR_3)_2(TCNQ)_2$$

R$_3$= Ph$_3$ **2a**, Ph$_2$Me **2b**, PhMe$_2$ **2c**

$$[Au(CH_2PR_3)_2]ClO_4 + LiTCNQ + TCNQ \longrightarrow Au(CH_2PR_3)_2(TCNQ)_2$$

Complex Au(CH$_2$PPh$_3$)$_2$(TCNQ)$_2$ (**1a**) crystallized in the triclinic space group P$\bar{1}$ and contains two Au(CH$_2$PPh$_3$)$_2$ units and two pairs of centrosymetrically related TCNQ molecules. The bond lengths and angles for the cation are in agreement with those found in other ylide complexes. The crystal lattice is not formed by segregated stacks of Au(CH$_2$PPh$_3$)$_2$ (D) and TCNQ (A) entities , but mixed stacks of donor and aceptor are formed, i.e. DADAAD which do not coincide closely with any of the crystallographic axes. The bond lengths and angles in the TCNQ molecules correspond with the presence of TCNQ⁻and TCNQ°·

In a similar procedure complexes [Au(PR$_3$)$_2$]ClO$_4$ react with LiTCNQ affording Au(PR$_3$)$_2$TCNQ(R$_3$= Ph$_3$ **3a**, Ph$_2$Me **3b**, (OMeC$_6$H$_5$)$_3$ **3c**, (MeC$_6$H$_5$)$_3$ **3d**) We obtain the same results carrying out the reaction either with LiTCNQ or with a mixture of LiTCNQ and TCNQ. Similarly, complexes **3a-d** do not react with an excess of TCNQ.The σ$_{RT}$ for complexes **2b** and **3a** are 2.3 10^{-6} S. cm^{-1} and 10^{-9} S. cm^{-1} respectively.

[Au(PPh$_3$)$_2$TCNQ] crystallized in the space group C2 with four molecules in the unit cell. The geometry around the gold atom is slightly distorted from linear with a P-Au-P angle of 174.7(3)º and showing a short contact of 3.207 Å with one N of the TCNQ. The crystal lattice is formed by two different segregated stacks of Au(PPh$_3$)$_2$ and TCNQ parallel to the x axis. The separation between the TCNQ$^-$ planes is 11.1 Å, but the separation between the TCNQ$^-$ of the two different stacks is 3.763 Å.

Other cationic gold complexes as [Au$_3$(CH$_2$PPh$_2$CH$_2$)$_2$ (PPh$_3$)$_2$]ClO$_4$ afford TCNQ derivatives by reaction with LiTCNQ .

[Au$_3$(CH$_2$PPh$_2$CH$_2$)$_2$(PPh$_3$)$_2$]ClO$_4$ + LiTCNQ ⟶

[Au$_3$(CH$_2$PPh$_2$CH$_2$)$_2$(PPh$_3$)$_2$]TCNQ

[Au$_3$(CH$_2$PPh$_2$CH$_2$)$_2$(PPh$_3$)$_2$]TCNQ crystallized in the space group P$\bar{1}$ with two molecules per unit cell. The distances between the TCNQ nitrogens and gold atoms are larger than expected for any short contact. This, and the fact that the TCNQ bond lengths are close to those found in RbTCNQ show that the TCNQ moiety is acting only as the counterion.

We have been using another approach to the synthesis of charge transfer complexes. Crystals of trans-diphenyltetrathiafulvalene (Ph$_2$TTF) complexes with bis(pentafluorophenyl)aurate(I) were grown by electrocrystallization from a 1,1,2-trichloroethane solution of Ph$_2$TTF (4.10^{-3}M) and NBu$_4$[Au(C$_6$F$_5$)$_2$] (0.01M) by using the crystal growth cell described elsewhere[4]. The constant current of 1.2 μA was applied (at 20° C and under N$_2$ atmosphere) for a period of 14 days. Black needles of (Ph$_2$TTF)$_2$[Au(C$_6$F$_5$)$_2$] (**5a**) were recovered. Working in a similar manner but in an acetonitrile solution of TTF (4.10^{-3} M) and NBu$_4$[Au(C$_6$F$_5$)$_2$] (0.1M) affords black prisms of (TTF)$_2$[Au(C$_6$F$_5$)$_2$] (**5b**). The σ_{RT} are 1.5 and 3 S.cm^{-1} for complexes **5a** and **5b** respectively.

The crystal structure of (Ph$_2$TTF)$_2$[Au(C$_6$F$_5$)$_2$] consists of two segregated stacks of Ph$_2$TTF and Au(C$_6$F$_5$)$_2$ with short contacts of 3.253 and 3.490 Å between one sulfur atom from each Ph$_2$TTF and the gold center. The separation between two Ph$_2$TTF molecules is 3.886 Å . The Au(C$_6$F$_5$)$_2$ moieties have very similar geometry to the free anion[5].

Acknowledgements

We thank the Fonds der Chemischen Industrie and The Dirección General de Investigación Científica y Técnica (n$^{\underline{o}}$ MAT90-0803) for financial support.

References

1   M. Ward, P.G. Fagan, J.C. Calabrese and D.C. Johnson, J Am. Chem. Soc., 1989, 111, 1719 and references therein.
2   T.J. Marks, Angew Chem.Int. Ed., 1990, 857.
3   V. Briois, R. M. Leguan, C. Cartier, G. van der Lean, A. Michalowicz and M. Verdaguer, Chem. Mater., 1992, 4,484.
4   A. Anzai, J. Moriya, K. Nozaki, J. Ukachi and G. Saito, J. Physique, 1983, 44, C3-1195.
5   R. Uson, A. Laguna, J. Vicente, J. Garcia, P.G. Jones and G.M. Sheldrick, J. Chem. Soc., Dalton Trans.,1981, 655.

Polynuclear Derivatives of Gold; Synthesis, Structure, and Reactivity

A. Laguna[1,*], M. Laguna[1], M. C. Gimeno[1], E. J. Fernández[1],
J. Jiménez[1], M. Bardají[1], J. M. López-de-Luzuriaga[1],
and P. G. Jones[2]

[1] DEPARTAMENTO DE QUÍMICA INORGÁNICA, INSTITUTO DE CIENCIA
DE MATERIALES DE ARAGÓN, UNIVERSIDAD DE ZARAGOZA, CSIC
50009 ZARAGOZA, SPAIN
[2] INSTITUT FÜR ANORGANISCHE UND ANALYTISCHE CHEMIE DER
TECHNISCHEN UNIVERSITÄT, HAGENRING 30, W-3300
BRAUNSCHWEIG, GERMANY

INTRODUCTION

The synthesis of polynuclear gold, silver or copper complexes has attracted considerable attention[1,2] in recent years and a variety of methods have been used to prepare these types of compounds.

Here we report the synthesis of polynuclear gold derivatives by using: **a)** the bis(ylide)gold(I) complex, $[Au(CH_2PPh_2CH_2)]_2$, which should possess an excess of electron density at the gold atoms, and **b)** the polydentate bis(diphenylphosphine)methanide or methandiide ligands.

1. Reactions with $[Au(CH_2PPh_2CH_2)]_2$

We have recently reported the preparation[3] of $[\{Au(CH_2PPh_2CH_2)\}_2Au(C_6F_5)_3]$, the first oligonuclear gold complex with a gold(I)-gold(III) bond unsupported by covalent bidges, obtained by the reaction of $[Au(CH_2PPh_2CH_2)]_2$ and $Au(C_6F_5)_3(tht)$, in which the bis(ylide)digold(I) should act as Lewis base.

The tetrahydrothiophene (tht) can also be readily displaced from asymmetrical gold(II) complexes to give hexanuclear gold derivatives (**1a,b,c**)

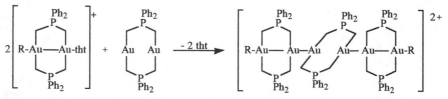

R= C_6F_5 (**1a**), $C_6F_3H_2$ (**1b**), Me (**1c**)

The structure of **1b** has been established by X-ray diffraction. The molecule lies on a centre of symmetry;

its backbone is a linear chain of six gold atoms, all of them with square planar coordination. An interesting feature is the difference found in the gold-gold distances [2.654(7) Å, external, and 2.838(7) Å internal diauracycle], which points out that the oxidation states II and I, of the precursors are likely retained.

$[Au(CH_2PPh_2CH_2)]_2$ does not react with gold(I) derivatives, such as $[Au(PPh_3)_2]^+$ or $[Au(L-L)]_2^{n+}$, with formation of complexes with gold-gold bonds; instead ylide migration takes place, affording polynuclear complexes. These ylide transfer reactions are the only way either to dinuclear gold complexes with only one $CH_2PPh_2CH_2^-$ bridging ligand or to asymmetrical double bridged gold derivatives.

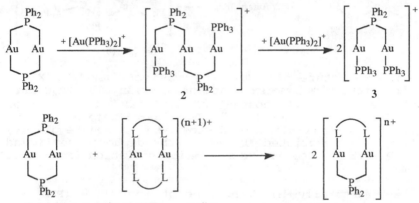

L-L= dppm, n=1 (**4a**); S$_2$CNEt$_2$, n= 0 (**4b**)

2. Reactions with (PPh$_2$)$_2$CH$^-$ or (PPh$_2$)$_2$C^{2-} ligands

The bis(diphenylphosphino)methanide (dppm-H) ligand has been studied as a versatile ligand in coordination chemistry. Usually, it acts as a bidentate or tridentate ligand by bonding to one, two or three metal atoms through the phosphorus and the central carbon atom.[4]

However, the eight-electron donor bis(diphenylphosphino)methandiide ligand has been scarcely studied and only three examples have been reported.[5]

We have found a route to such complexes by means of acetylacetonate derivatives. This has allowed us to synthesize different polynuclear gold derivatives.

The bis(diphenylphosphino)methanide complex [AuR$_2$-{(PPh$_2$)$_2$CH}] possesses an excess of electron density on the methanide carbon, causing it to act as a C-donor nucleophile. Therefore it can displace weakly coordinated ligands such as tetrahydrothiophene from

[Au(tht)PPh₃]ClO₄, forming the binuclear complex [AuR₂-{(PPh₂)₂CH(AuPPh₃)}], **5**.

The reaction of **5** with [Au(acac)PPh₃] or [N(PPh₃)₂]-[Au(acac)Cl] leads to proton substitution by the fragments AuPPh₃⁺ and AuCl and formation of acetylacetone, giving the trinuclear complexes [AuR₂-{(PPh₂)₂C(AuPPh₃)₂}]ClO₄, **6** or [AuR₂{(PPh₂)₂C(AuPPh₃)-(AuCl)}], **7**. Compound **6** can also be prepared by direct reaction of the diphosphine complex [AuR₂-{(PPh₂)₂CH₂}]ClO₄ with two equivalents of [Au(acac)PPh₃].

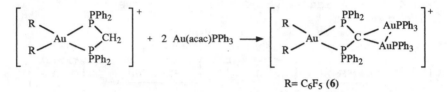

R= C₆F₅ (**6**)

The structure of complex **6** was confirmed by X-ray diffraction. The complex contains a triangular Au₂C unit with a short Au...Au contact of 2.826(2) Å and a narrow Au-C-Au angle of 85.4(7)°. The coordination of the gold(III) atom is slightly distorted from square planar, whereby the restricted "bite" of the diphosphine ligand P-Au-P 70.5(3)° represents the major diviation from ideal geometry.

A structurally different type of complex, [N(PPh₃)₂]-[{AuR₂{(PPh₂)₂C(AuR)}}₂Au] **8**, is obtained when [AuR₂-{(PPh₂)₂CH(AuR)}] is treated with a half-equivalent of [N(PPh₃)₂][Au(acac)₂], which has the ability to abstract two protons and thus to afford higher nuclearity derivatives.

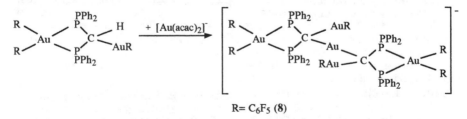

R= C₆F₅ (**8**)

Compounds **5-8** have been characterized by analytical, IR and NMR spectroscopy, and FAB mass spectrometry.

We have extended our studies in this type of system to the bis(phosphine)sulphide ligand CH₂(Ph₂PS)₂ by reacting the latter with [Au(acac)PPh₃]. The reaction leads to the complex (SPPh₂)₂C(AuPPh₃)₂ **9** which is the first example with (SPPh₂)₂C²⁻, potentially an eight-electron donor ligand. The sulfur atom donors can be used to coordinate further to a metal centre. Therefore, complex **9** reacts with two equivalents of [Au(C$_6$F$_5$)tht] affording the pentanuclear complex [Au$_5$(C$_6$F$_5$)-{(SPPh$_2$)$_2$C}$_2$(PPh$_3$)], **10**.

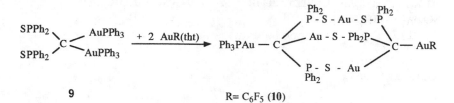

9   R= C$_6$F$_5$ **(10)**

The structure of **10** was confirmed by X-ray diffraction. The coordination of the five gold(I) atoms is linear, with a maximun deviation from linearity of 13°. It is worth mentioning the short Au...Au contacts that lie in the range of 2.988(6)-3.337(6) Å.

Acknowledgements

We thank the Dirección General de Investigación Científica y Técnica (no. PB88 0075) and the Fonds der Chemischen Industrie for financial support.

REFERENCES

1   D.M.P. Mingos and M.J. Watson, <u>Transition. Met. Chem.</u>,1991, <u>16</u>, 285.

2   B.K. Teo and H. Zhang, <u>Angew. Chem. Int. Ed. Engl.</u>, 1992, <u>31</u>, 445.

3   R. Usón, A. Laguna, M. Laguna, M.T. Tarton and P.G. Jones, <u>J. Chem. Soc., Chem. Commun.</u>, 1988, 740, and references cited therein.

4   B. Chaudret, B. Delavaux and R. Poilblanc, <u>Coord. Chem. Rev.</u>,1988, <u>86</u>, 191, and references cited therein.

5   V. Riera and J. Ruiz, <u>J. Organomet. Chem.</u>, 1986, <u>310</u>, C36; M. Luser and P. Peringer, <u>Organometallics</u>, 1984, <u>3</u>, 1916; S.I. Al-Resayes, P.B. Hitchcock and J.F. Nixon, <u>J. Chem. Soc., Chem. Commun.</u>, 1986, 1710.

New Carbene Complexes of Gold and Copper

H. G. Raubenheimer, R. Otte, and S. Cronje

DEPARTMENT OF CHEMISTRY AND BIOCHEMISTRY, RAND
AFRIKAANS UNIVERSITY, PO BOX 524, JOHANNESBURG 2000,
SOUTH AFRICA

1 INTRODUCTION

Synthetic pathways towards isolable carbene complexes are few for gold[1],[2],[3] and non-existent for copper. Carbene complexes of copper have, however, been postulated as intermediates in a variety of reactions.[4]

We have set out to develop a method of carbene complex synthesis based on thiazolyllithium, $LiC=NC(R^1)=C(R^2)S$ (R^1 = Me, R^2 = H; $R^1 R^2$ = $C_4 H_4$), and various metal chlorides as starting materials. The results obtained are related to the recent work of Burini and co-workers who prepared gold derivatives of imidazoles.[5]

2 RESULTS AND DISCUSSION

Lithiation of benzothiazole (giving a-compounds) and 4-methylthiazole (giving b-compounds) with LiBu at -70 °C and subsequent addition of one half mole equivalent of [ClAu(tht)] (tht = tetrahydrothiophene) afforded corresponding aurate complexes, one of which (1) was isolated and structurally characterized as its tetrabutylammonium salt (Scheme 1). The 'ate' complexes were also directly protonated (HCl in ether or $CF_3 SO_3 H$) and alkylated ($CF_3 SO_3 Me$) to afford neutral monocarbene (2,3) and cationic bis(carbene) compounds (4,5). The initial consecutive addition of two different lithiated thiazoles failed to produce mixed complexes, indicating the preference of gold(I) towards the formation of homoleptic aurates. A complication occurred during the attempted preparation of (3b), when alkylation occurred on the metal-bonded carbon atom resulting in (6). The cationic carbene(thiazole) complex (7) was also characterized.

The crystals of (2b) contain dimeric complex entities in the unit cell, with a gold-gold separation of only 3.07Å. Hydrogen bonding which occurs between the nitrogen atoms in the dimer of crystalline (2b) is,

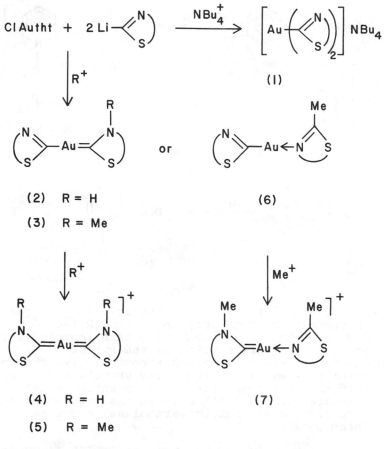

Scheme 1

however, not a prerequisite for gold-gold bond formation, and in the cation (4b), where hydrogen bonding is
absent, a distance of 3.33Å between gold atoms in the
dimer was obtained by X-ray crystallography. All the
gold-carbene distances determined in this study vary
between 1.95 and 2.06Å. Before final purification, white
microcrystals of (2b) converted spontaneously into
bright-red, extremely disordered crystals with a hexagonal
structure in which the gold atoms are arranged in linear
chains 3.0Å apart. Later attempts to prepare more of the
red crystals failed.

Mono(carbene) complexes were also prepared with
other neutral [PPh₃, (8,9)] and anionic [CN⁻, (10,11);
C₆F₅, (12,13)] ligands attached to gold, by respectively
protonating or alkylating the substitution products.
Several complicating reactions occurred during these
transformations: (i) the intermediate
[PPh₃Au(C=NC(Me)=CHS)] spontaneously converted at room
temperature into the polymer [AuC=NC(Me)=CHS]ₙ (14), (n

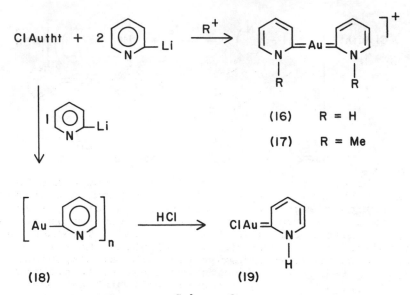

Scheme 2

probably 3, vide infra); (ii) alkylation of the reaction
mixture containing [PPh₃AuCl] and LiC̄=NC(R¹)=C(R²)S̄
produced only the homoleptic bis(carbene) complexes (5a)
and (5b) and (iii) protonation of the aurate
[C₆F₅Au(C̄=NC(Me)=CHS̄)]⁻ yielded a neutral pentafluoro-
(thiazole) complex (15). The X-ray structure of (8b)
indicated that the complexes occur as antiparallel pairs
(Au-Au separation ca. 3.1Å) whereas the gold atoms in
(13b) are 3.95Å apart but, nevertheless, occur as
non-linear chains.

The sulphur atoms present in the thiazole CH acids
are not necessary for preparing carbene complexes
according to our ate-method. Lithiated pyridine also gave
pyridinylidene compounds (Scheme 2). Our efforts to
prepare neutral pyridyl-pyridinylidene compounds were
unsuccessful and only the cationic carbene complexes (16)
and (17) could be isolated. Gold-gold interactions in the
crystals in (16) are prevented by hydrogen bonding to the
counter ion (Cl⁻) and to a water molecule. The proposed
trimer (18)[6] was acidified to give a carbene(chloro)
complex (19) which might be of further use in the
preparation of mixed carbene complexes.

Lithiated thiazoles have also been used to prepare
carbene complexes of other metals such as chromium,[7] iron[8]
and, very recently, platinum.[9] Treatment of [Cl₂Pt(tht)₂]
with benzothiazolyllithium followed by protonation with
CF₃SO₃H afforded only a dicationic tetracarbene complex.
A crystal structure determination showed that its
structure to a large extent resembles that of
[Pd{C̄N(H)CH₂CH₂Ō}₄]Cl₂, which was obtained by a completely
different route.[10]

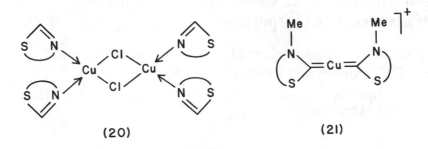

(20)   (21)

The copper analogues of (2) or (4) could not be obtained by acidification of the corresponding cuprates. Protonation occurred on the co-ordinated carbon, leading for example to compounds of type (20) which could not be prepared directly from CuCl and thiazole. Treatment of the thiazolyl ate compounds with CF_3SO_3Me, however, produced two desired cationic bis(carbene) compounds, (21). Up to now we have been unsuccessful in growing crystals of these novel but relatively stable copper carbene complexes suitable for X-ray study. Their structures were assigned on the basis of elemental analyses, mass spectra [m/z corresponding to the cationic complex ion (21a), the appearance of various expected fragments of (21b)] and ^1H as well as ^{13}C NMR spectra [NMe at δ 4.39 and 4.12 ppm and CuC at δ 216.3 and 212.5 ppm for (21a) and (21b) respectively].

REFERENCES

1.   B. Cetinkaya, P. Dixneuf and M. Lappert, *J. Chem. Soc., Dalton Trans.*, 1974, 1827.
2.   G. Minghetti, L. Baratto and F. Bonati, *J. Organomet. Chem.*, 1975, 102, 397.
3.   R. Aumann and E.O. Fischer, *Chem. Ber.*, 1981, 114, 1853.
4.   W. Carruthers in 'Comprehensive Organometallic Chemistry', ed. G. Wilkinson, F.G.A. Stone and E.W. Abel, Pergamon Press, Oxford, 1982, p.661.
5.   F. Bonati, A. Burini, B.R. Pietroni and B. Bovio, *J. Organomet. Chem.*, 1989, 375, 147.
6.   L.G. Vaughan, *J. Am. Chem. Soc.*, 1970, 92, . 730.
7.   H.G. Raubenheimer, G.J. Kruger, A. van A. Lombard, L. Linford and J.C. Viljoen, *Organometallics*, 1985, 3, 275.
8.   H.G. Raubenheimer, F. Scott, S. Cronje, P.H. van Rooyen and K. Psotta, *J. Chem. Soc., Dalton Trans.*, 1992, 1009.
9.   H.G. Raubenheimer and W. van Zyl, unpublished results.
10.   W.P. Fehlhammer, K. Bartel, U. Plaia, A. Völkl and A.T. Liu, *Chem. Ber.*, 1985, 118, 2235.

New Multi-metallic Compounds: Syntheses, Structures, and Properties

Alexander J. Blake, Robert O. Gould, Graig M. Grant,
Paul E. Y. Milne, and Richard E. P. Winpenny*
DEPARTMENT OF CHEMISTRY, THE UNVERSITY OF EDINBURGH,
EDINBURGH EH9 3JJ, UK

Introduction

Several groups are currently synthesising and studying mixed d-block/f-block metal complexes.[1] Our approach has used the ambidentate bridging ligand 2-pyridone (LH), and the 6-substituted derivatives thereof. An extensive series of mixed copper/lanthanoid complexes have been prepared and structurally characterised.[2-4] The stochiometry and structure of the product depends on the ligand used, and on reaction conditions. More recently, this work has been extended to include complexes of copper with s-block metals.

Cu-Ln Complexes of 2-Pyridone (LH)[2]

The mixed-metal complexes so far characterised for this ligand appear to depend largely on the anion present, and also on the particular lanthanoid used. Addition of hydrated lanthanum perchlorate to a methanol solution of hydrated copper nitrate and the deprotonated ligand led to the isolation of a La_4Cu_4 complex, the core of which is shown in Figure 1a. This complex consists of two copper dimers which are separated by a plane of four lanthanum ions. The bridging within the compound is extremely complicated, involving trinucleating hydroxides, methoxides, and deprotonated ligands. If a perchlorate salt of a later lanthanoid is used, a purple powder is obtained, which on treatment with HCl or HBr reacts to give a quite different structural type (Figure 1b). This contains a Ln_2Cu_4 core, with the six metals arranged in an octahedral fashion; the lanthanoid atoms at the apices, and the copper atoms in the equatorial plane. Magnetic measurements on the Gd_2Cu_4 complex reveal weak ferromagnetic coupling between the Cu^{2+} and Gd^{3+} ions.

Cu-Ln Complexes of 6-Chloro-2-Pyridone (ClLH)[3]

Using the 6-chloro derivative of the ligand, a copper dimer of formula $Cu_2(ClL)_4$ could be isolated (Figure 2), which has a copper acetate like structure. The Cl substituent prevents any axial ligation, and this is perhaps reflected in a shorter than usual Cu-Cu bond of

Figure 1.
(a) The core of an La$_4$Cu$_4$ complex, with atoms not involved in metal-metal bridging omitted for clarity.

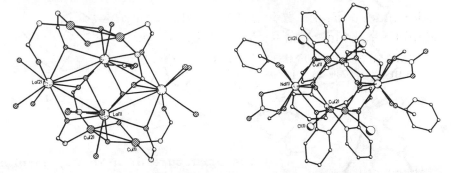

(b) The structure of a Cu$_4$Nd$_2$ complex.

2.4989(11)A, cf 2.56 A for a dimer with 3-methyl-2-pyridone which has DMF ligands in axial positions.[5] Crystals and dichloromethane solutions of Cu$_2$(ClL)$_4$ are deep red. Preliminary electrochemical measurements show two irreversible reductions at -0.21 and -0.88 V. The magnetic moment at room temperature is 1.58 BM for the dimer, indicating anti-ferromagnetic coupling between the coppers.

Figure 2. The structure of Cu$_2$(ClL)$_4$

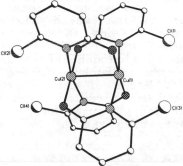

The dimer reacts with hydrated lanthanoid nitrates in methanol to give new Ln$_x$Cu$_y$ complexes. Two have been structurally characterised. The first is with lanthanum, and is an asymmetrical LaCu$_3$ complex (Figure 3a). A residual dimer like structure can perhaps be seen in the region of Cu(2) and Cu(3). The short metal-metal contacts are La(1)-Cu(2) 3.719, Cu(1)-Cu(3) 3.108 and Cu(2)-Cu(3) 2.970 A. With ytterbium nitrate a completely different structure was found. This is a centrosymmetric Yb$_2$Cu$_2$ complex (Figure 3b). At the centre of the molecule is a Cu$_2$O$_2$ ring, with the oxygens coming from deprotonated methanol. The coppers are four coordinate, with the remaining sites occupied by N atoms from ClL ligands. These ligands then bridge to the Yb atoms, each of which is eight coordinate. The Cu-Cu contact is 2.94 A, and the closest Yb-Cu contact is 4.27 A.

Figure 3. (a) The structure of a LaCu$_3$ complex.

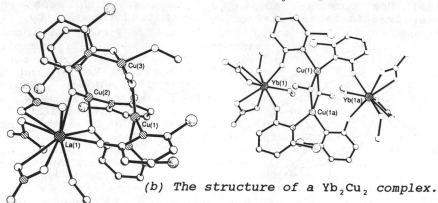

(b) The structure of a Yb$_2$Cu$_2$ complex.

Cu-Ln Complexes of 6-Methyl-2-Pyridone (MeLH) [4]

With the 6-methyl derivative a hexanuclear copper complex of formula [Cu$_6$Na(MeL)$_{12}$][NO$_3$] was isolated and structurally characterised. It is a "metallocrown" with at its centre a six coordinate sodium ion. There are two types of copper site. In each the copper is four coordinate, with bonds to two nitrogens and two oxygens from MeL units (Figure 4), however the arrangement of donors is *cis* for one copper site, and *trans* for the second. The molecule retains this structure in CHCl$_3$, as shown by nmr where six aromatic resonances are seen between 41 and 18 ppm, and two methyl resonances are observed at -7 and -14 ppm. This shows there are two types of ligand being paramagnetically shifted, in accord with the structure where again there are two ligand types. FABMS of the compound gave strong peaks for the molecular formula, and for fragments Cu$_4$Na(MeL)$_8$, Cu$_3$Na(MeL)$_6$, Cu$_2$Na(MeL)$_4$, Cu$_2$Na(MeL)$_3$ and CuNa(MeL)$_2$ – illustrating the Na ion is very firmly held within the cavity.

Figure 4. The structure of the cation [Cu$_6$Na(MeL)$_{12}$]$^+$

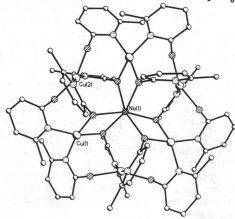

Reaction of the hexanuclear copper complex with hydrated lanthanum nitrates in methanol gives a series of Ln_2Cu_2 complexes (Ln = La, Ce, Nd, Sm, Gd, Dy and Yb) of similar structure to that shown for Yb_2Cu_2 in Figure 3b above. Preliminary magnetic measurements on the Gd_2Cu_2 complex indicate anti-ferromagnetic coupling of the copper ions, and weak ferromagnetic coupling between the Cu and Gd ions. In the absence of methanol a novel complex containing a La_8Cu_{12} core resulted. The core is shown in Figure 5. It consists of a cube of La atoms; at the centre of each of the cubes edges is a Cu atom, thus forming a cuboctahedron of coppers. The array is held together by twenty-four μ_3-OH ions, and has non-crystallographic O_h symmetry. The coordination of the La atoms is completed by bidentate nitrates, and MeLH groups (not shown). At the centre of the complex is a nitrate ion. The cube edges average 6.57 Å, and the close metal-metal contacts are Cu-Cu, average 3.37 Å and Cu-La 3.53 Å.

Figure 5. The La_8Cu_{12} array bridged by twenty-four μ_3-hydroxides.

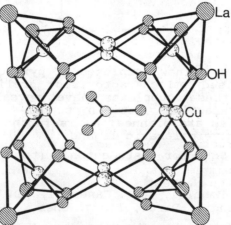

Reactions of $[Cu_6Na(MeL)_{12}]^+$ with other metals

The copper metallocrown reacts with $Pb(NO_3)_2$ to give a new complex in which the sodium ion is displaced by another copper, giving $[Cu_6Cu(MeL)_{12}][Pb(NO_3)_4]$ (Figure 6a). The cation is almost superimposable on $[Cu_6Na(MeL)_{12}]$. The mechanism of this reaction is unclear. FABMS gives peaks for the cation, plus fragments $Cu_4Cu(MeL)_8$, $Cu_4Cu(MeL)_7$, $Cu_3Cu(MeL)_7$, $Cu_3Cu(MeL)_6$ and $Cu_3Cu(MeL)_5$ - very similar to the fragmentation observed for $[Cu_6Na(MeL)_{12}]$. Additional peaks are seen above the mass of the cation, which are correct for $Cu_6Cu(MeL)_{12}(NO_3)$, $Cu_6Cu(MeL)_{12}(NO_3)Pb$, $Cu_6Cu(MeL)_{12}$ $(NO_3)_2Pb$ and $Cu_6Cu(MeL)_{12}(NO_3)_3Pb$. With $Mg(NO_3)_2$ again the sodium is replaced, but now the metallocrown is broken in half to give a Cu_3Mg complex (Figure 6b) in low yield. With this reaction the major product was unreacted $[Cu_6Na(MeL)_{12}][NO_3]$.

Figure 6.
(a) The structure of the dication $[Cu_6Cu(MeL)_{12}]^{2+}$

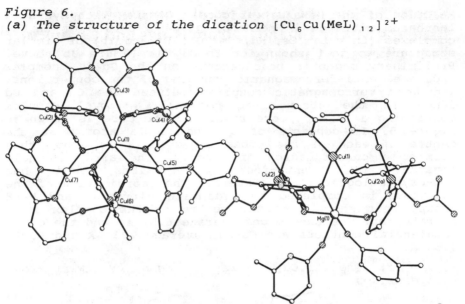

(b) The structure of a Cu_3Mg *complex*

Conclusions

A large number of new complexes of pyridone ligands have been synthesised and characterised; preliminary magnetic studies of these complexes have revealed interesting magnetic interactions between copper and lanthanoids. Further studies are planned, as is further synthetic work using other heteroaromatic ligands.

We thank the SERC for a studentship (to P.E.Y.M.).

References

1. (a) C. Benelli, A. Caneschi, D. Gatteschi, O. Guillou and L. Pardi, *Inorg. Chem.*, 1990, **29**, 1751.

    (b) N. Matsumato, M. Sakamoto, H. Tamaki, H. Okawa and S. Kida, *Chem. Lett.*, 1989, 853.

    (c) O. Guillou, R. L. Oushoorn, O. Kahn, K. Boubekeur and P. Batail, *Angew. Chem., Int. Ed. Engl.*, 1992, **31**, 626.

2. A. J. Blake, P. E. Y. Milne, P. Thornton and R. E. P. Winpenny, *Angew. Chem., Int. Ed. Engl.*, 1991, **30**, 1139.

3. A. J. Blake, R. O. Gould, P. E. Y. Milne and R. E. P. Winpenny, *J. Chem. Soc., Chem. Comm.*, 1992, 522.

4. A. J. Blake, R. O. Gould, P. E. Y. Milne and R. E. P. Winpenny, *J. Chem. Soc., Chem. Comm.*, 1991, 1453.

5. Y. Nishida and S. Kida, *Bull. Chem. Soc. Jpn.*, 1985, **58**, 383.

Small Clusters of Copper Triad Compounds with Bridging Sulfur Ligands – from Rings and Chains to Butterflies and Cubanes

John P. Fackler, Jr., A. Elduque, R. Davila,
David (C.W.) Liu, C. Lopez, T. Grant, C. McNeal,
R. Staples, and T. Carlson

LABORATORY FOR MOLECULAR STRUCTURE AND BONDING AND
THE DEPARTMENT OF CHEMISTRY, TEXAS A&M UNIVERSITY,
COLLEGE STATION, 77843, TX, USA

Introduction. Since 1960, a substantial portion of our work has been with compounds of Group 11 elements.[1] This paper addresses the most recent progress we have made with the chemistry of Group 11 element compounds. It includes oxidative addition involving gold and related elements,[2-8] especially the new results of Fenske-Hall calculations on linear trinuclear species. Work with gold thiols, chemistry that is related to the use of gold drugs in treatment of rheumatoid arthritis,[9] has led us to some new observations regarding the coordination of gold to dithiolene sulfur ligands. Olefin bonding also is observed and many of these compounds are strongly luminescent, an observation which may be related to singlet oxygen quenching by Au(I) in diseased cells. Dinuclear Au(I) complexes with Au...Au interactions often display phosphorescence. Certain mononuclear species also luminesce. This observation is described. Polynuclear copper(I) and silver(I) complexes with sulfur ligands again have attracted our interest, especially the dithiolate and sulfur-rich dithiolate tetrahedranes and homocubanes. We desire to better understand the bonding and electronic properties of these compounds and to develop rational procedures to synthesize mixed metal clusters.

Trinuclear oxidative addition. The addition of halogens to linear trinuclear, M..M'..M, complexes[2] (Sketch 1) can produce two metal-metal bonds if the metal atom in the middle, M', can contribute two electrons and two orbitals to the process. The reaction is an "addition" in the Goddard-Low[10] sense, requiring a promotion of the atoms from closed shell or pseudo closed shell configurations to open shell configurations. With M' = Pt(II), this readily happens and **1** reacts to form **4**. M' = Pb(II), **2**, no metal-metal bonded product can be isolated. Fenske-Hall calculations[11] suggest that the electron promotion required for oxidative addition comes at the expense of antibonding metal-metal interactions with Pt. With **2**, Au..Pb bonding orbitals contribute strongly to the HOMO and SHOMO. The

Pb compound decomposes in the oxidation process. For the newly synthesized Hg(II) complex, **3**, isoelectronic with **1**, no reaction other than decomposition is observed. Electron promotion here is from a HOMO that involves no metal-metal interactions. The HOMO - LUMO gap also is quite large, 6.81 eV.

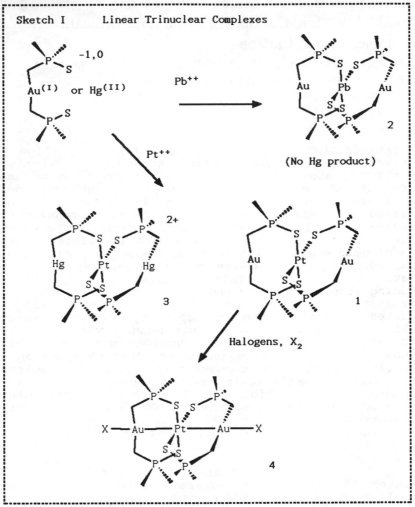

Sketch I   Linear Trinuclear Complexes

(No Hg product)

Halogens, X₂

Dinuclear dithiolates. Recently[12] we described the first dianionic dithiolates formed with dinuclear Au(I). Unlike the well known dithiocarbamates[13] and related neutral species, these materials oxidize to form isolable metal-metal bonded dinuclear compounds. The optical properties of these compounds, with their polarizable Au-S bonds, are potentially of interest in non-linear optics.[14] Therefore we have embarked upon a general study of the chemistry and optical properties of dinuclear gold(I) dithiolates. To date we have synthesized several ring compounds such as [Au$_2$(μ-dppm)(S(CH$_2$)$_3$S)], **5**, which show short Au..Au distances and visible phosphorescence. In general in these dinuclear gold(I) compounds we believe that emission arises from excitation of the largely σ*(s,d$_{z^2}$,d$_{x^2-y^2}$) to σ(p$_z$) with formation of a triplet state. This triplet state may involve the p$_z$ orbital or the p$_y$ orbital which is calculated[15] (less reliable for empty orbitals) by Xα-SCF to be at a somewhat lower energy than p$_z$(Sketch 2). Recently it has become

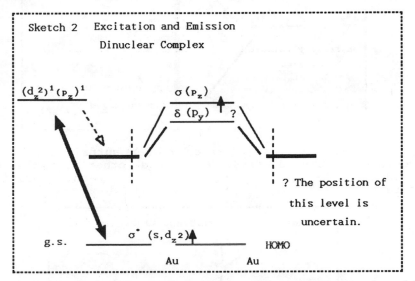

Sketch 2   Excitation and Emission
            Dinuclear Complex

$(d_z{}^2)^1(p_z)^1$ σ(p$_z$)

                             δ(p$_y$)   ?

? The position of
this level is
uncertain.

g.s. σ*(s,d$_z{}^2$)   HOMO

            Au   Au

established that some three coordinate mononuclear[16] gold(I) complexes luminesce in the solid state and in solution. It is thought that this emission arises from a triplet state involving the p$_z$ orbital perpendicular to the three coordinate metal-ligand plane containing the (d$_{x^2-y^2}$, d$_{xy}$) orbitals. (The orientation of this orbital is the same as the δ orbitals in the dinuclear complexes.) Excitation to this level occurs through the spin allowed transition from d$_{z^2}$ followed by non-radiative crossover (Sketch 3), or directly to this level by virtue of spin-orbit coupling.

The 1,2-dithiolene ligands bond to triphenylphosphine gold(I) differently from the 1,1-dithiolenes. The ligand appears to reorganize electronically into a coordinated thiol and a coordinated "allyl", with the latter sulfur ligand contributing less than two electrons to the coordination. This gold atom bridges over to the thiol sulfur atom as a result. We have prepared several other bis-(triphenylphosphine)gold-1,2-dithiolene species and find a similar coordination behavior. The

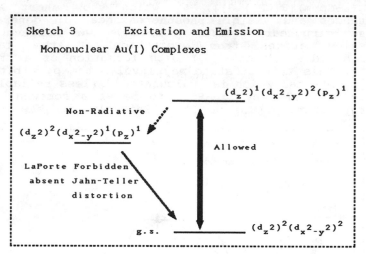

[1,2-dithiolenegold(I)]⁻ anion rather surprisingly also coordinates to olefins. Only one other example of gold(I) olefin coordination has been verified structurally.[8]

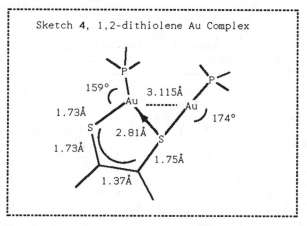

Small Clusters. We recently reported that mercapto-thiazoline, MT, complexes of copper(I) are tetranuclear "butterfly" complexes[5]. The compounds contain trigonally and tetrahedrally coordinated Cu(I), and in the absence of coordinating ligands are polymerized clusters of clusters. Bases coordinate to the "wing-tip" atoms. The silver(I) clusters appear to be isostructural. In solution, a "hinge" - "wing-tip" interconversion appears to occur to make all four MT ligands equivalent. The symmetric intermediate is a Cu_4 tetrahedron. With both

metal atoms present, a mixed metal cluster is obtained
with silver(I) atoms at the "wing-tips". ^{31}P NMR studies

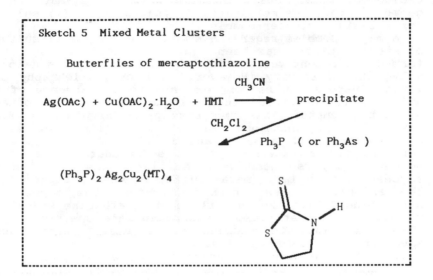

Sketch 5 Mixed Metal Clusters

Butterflies of mercaptothiazoline

$$Ag(OAc) + Cu(OAC)_2 \cdot H_2O + HMT \xrightarrow{CH_3CN} precipitate$$

$$\xleftarrow[Ph_3P \ (\ or \ Ph_3As \ )]{CH_2Cl_2}$$

$(Ph_3P)_2 \, Ag_2Cu_2(MT)_4$

in the solid state on these materials show [31]P-[109,107]Ag
coupling consistent with the Ag-P bond lengths[17]
observed for the complexes. Two lines with J(Hz) = 379
are observed for the mixed metal complex while only one
broad line is found for the tetranuclear species. The
copper complex shows four lines due to the spin of 3/2
for the Cu nucleus.

In 1968 we reported[6] the crystal structure of a
compound synthesized earlier in our group which displayed a homocubane arrangement of copper(I) atoms. This
was the first time a homocubane of metal atoms had been
observed. Since this time additional work has been done
in Coucouvanis's laboratory[18] on the Cu(I) system and
homocubane structures have been found with Ag, Ni,
Co,Fe, and Pd. Some metal centered homocubanes of the
type $M_9L_8L'_6$, M = Ni, Pd, L = CO, PR_3, L' = group 14,15,
or 16 ligands, also have been obtained.[19] A mixed metal
homocubane structure of Fe_3Ni_5 also was reported.[19b]
With group 11 homocubanes, there has been some question
concerning the metal-metal interactions present. Are the
compounds really clusters? With Avdeef,[20] we suggested
that "metal-metal bonding does not appear to be very
important...".

Recent work[21] has shown that a Cu_8S_{12} cluster
apparently exists in yeast metallothionein and a copper
thiolate cluster in the protein CUP2 that regulates the
expression of this enzyme shows[22] a Cu-Cu vector of 2.7
Å. Also Kanatzidis[23] has succeeded in characterizing a
$Cu_8Se_{12}Na$ solid state cluster containing material with
the Na in the middle of the homocubane. An old structure
also reports Cl to be in the center of a copper

dithiophosphate cubane.[24] In all of these compounds the copper(I) atom remains essentially three coordinate excluding metal-metal interactions. A planar CuS_3 geometry about the metal atom dominates copper(I) sulfur (or selenium) cluster chemistry.

A major problem regarding studies of copper(I) sulfur cluster chemistry has been characterization of species formed. In recent years x-ray crystallography has helped solve the difficulty. However, even crystallography of large clusters can be complex, and in the absence of a quick characterization tool, structural studies often are not attempted. Mass spectroscopy in conjunction with crystallography has been of enormous help. Yet mass spectroscopy such as FABS and field desorption, while very helpful and even used by us to characterize the Ag_2Cu_2 clusters described earlier, has not been successful with large metal-sulfur clusters, especially the anionic species found to produce the copper(I) homocubanes. With the support of Dr. Catherine McNeal,[7] we have found that plasma desorption mass spectroscopy, PDMS, is able to succeed where these other mass spectroscopic tools have failed.

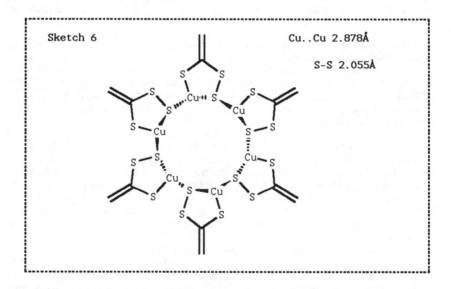

Sketch 6   Cu..Cu 2.878Å

                                  S-S 2.055Å

In an attempt to generalize some of the copper(I) cubane cluster chemistry reported by Coucouvanis[18] and to develop a rational synthesis of mixed metal homocubanes, we have combined PDMS studies with crystallography and have looked again at $Cu_8(i\text{-}MNT)_6^{4-}$, i-MNT = $S_2CC(CN)_2^{2-}$, and its reactions with protonic acids and sulfur. With the ligand DED, 1,1-dicarbo-*tert*-butoxy-ethylene-2,2-dithiolate, Coucouvanis had demonstrated that protonic acids form protonated ligand copper(I) cluster tetrahedranes, opening up the cube. This reaction suggested reversibility might be possible and potentially represents a way to synthesize mixed metal homocubanes. With sulfur, addition to form a sulfur-rich species appears to take place, $[Cu_8(S-Bu^t-DED)_6]^4$. This material oxidizes to form a square pyramidal Cu_5 species[25] containing four Cu(I) and one Cu(III), (Au(III) can replace Cu(III)).

PDMS studies of $[Cu_8(i-MNT)_6]^{4-}$, **5**, have detected the parent ion, $(cation)_5[Cu_8(i-MNT)_6]^+$, and a species $(cation)_6[Cu_8(i-MNT)_6]^+$ which remains uncharacterized structurally. Sulfur addition to **5** produces the structurally characterized anion $[Cu_6(S-i-MNT)_6]^{6-}$, (Sketch 6). In addition we have succeeded in structurally characterizing a $[Cu_4(i-MNT)_4]^{4-}$ tetrahedrane of copper(I) atoms from a tetrameric xanthate.

Acknowledgments. Hubert Schmidbaur[24] and his students, and Rafael Uson with Antonio Laguna[25] and their students have pioneered studies in gold chemistry that have stimulated our work in this area. I also appreciate receipt of a **Bye** Fellowship, Robinson College, Cambridge and the **1992 Wilhelm-Manchot Forschungprofessur** position at the Technischen Universitat Munchen. The U.S. National Science Foundation , CHE 8708625, and the Robert A. Welch Foundation have supported this work along with the Texas Advanced Research Program.

References.
1. John P. Fackler, Jr., F.A. Cotton, J. Chem. Soc.,1960, 1435.(Senior author's first publication with copper(II).) J. P. Fackler, Jr., Prog. Inorg. Chem., 1976, 21, 55. J. P. Fackler, Jr., L.C. Porter, J. Amer. Chem. Soc., 1986, 108, 2750.
2. H.H. Murray, D.A. Briggs, G. Garzon, R. Raptis, L. C.Porter, J.P. Fackler, Jr., Organometallics,1987,6, 1992; C. King, D.D. Heinrich, G. Garzon, J-C.Wang, J.P. Fackler, Jr.,J. Amer. Chem. Soc., 1989, 111, 2300.
3. J.P. Fackler, Jr.,C. Paparizos, J. Amer. Chem. Soc.,1977, 99, 2363.
4. J.P. Fackler, Jr., J. Basil, Organometallics, 1982, 1, 871; A.M. Mazany, J.P. Fackler, Jr.,J. Amer. Chem. Soc., 1984, 106, 801; J.P. Fackler, Jr., H.H. Murray, Organometallics, 1984,3, 821.
5. J.P. Fackler, Jr., C.A. Lopez, R. Staples, S. Wang, R.E.P. Winpenny, J. Chem. Soc., Chem. Commun.,1992,146.
6. L.E. McCandlish, E.C. Bissell, D. Coucouvanis, J.P. Fackler, Jr., K. Knox, J. Amer. Chem. Soc., 1968, 90, 7357; J.P. Fackler, Jr., D. Coucouvanis,J. Amer. Chem. Soc., 1966, 88, 3913.
7. J.P. Fackler, Jr., C.J. McNeal, R.E.P. Winpenny, L.H. Pignolet, J. Amer. Chem. Soc., 1989, 111, 6434.
8. D.B. Bell'Amico, F. Calderazzo, R. Datona, J. Strahle, H. Weiss,Organometallics, 1987,6, 1207.

9. S.C. Stinson,Chem. & Eng. News, Oct. 16, 1989, p. 37;R.V. Parish, S.M. Cottrill, Gold Bulletin, 1987,3.; R.C. Elder, K. Tepperman, Proc. First Conf. Gold and Silver in Medicine, Wash. D.C., May 13-14,1987.

10.J.J. Low, Wm. Goddard,III, J.Amer.Chem.Soc., 1984, 106, 6928; ibid., Organometallics, 1986,5, 609; ibid., J. Amer. Chem. Soc., 1986,108, 6115.

11.A. Sargent, M.B. Hall, T.F. Carson, J.P. Fackler, Jr., to be published.

12.Md.N.I. Khan, S. Wang, J.P. Fackler,Jr.,Inorg. Chem., 1989, 28, 3579 and references therin.

13.See reviews by D. Coucouvanis, Prog. Inorg. Chem.1970, 11, 233; ibid.,1979, 26, 301.

14.D.S. Chemla, J. Zyss,editors, "Optical Properties of Organic Molecules and Crystals",Academic Press, Orlando, FLA.,1987.

15.C. King, J-C. Wang, Md.N.I. Khan, J.P. Fackler, Jr., Inorg. Chem., 1989, 28, 2145.

16.C. King, Md.N.I. Khan, R.J. Staples, J.P. Fackler, Jr.,Inorg. Chem., 1992, in proof.; T.M. McClesky, H.B. Gray,Inorg.Chem., 1992, 31, 1733.

17.N.J. Claydon, Chemica Scripta, 1988, 21,211.

18.F.J. Hollander, D. Coucouvanis, J.Amer.Chem.Soc., 1974, 96 , 5646; ibid.,1977,99, 6268; S. Kanodia, D. Coucouvanis, Inorg.Chem., 1982, 11, 469; D. Coucouvanis, D. Swenson, N.C. Baenziger, R. Pedelty, M.L. Cafferty, S. Kanodia, Inorg.Chem., 1989, 28, 2829.

19.a) J.P. Zebrowski, R.K. Hayashi, A. Bjarnason, L.F. Dahl, J. Amer. Chem. Soc.,1992,114, 3121. b) W. Saak, S. Pohl, Angew. Chemie, Int. Ed. Eng.,1991, 881.

20.A. Avdeef, J.P. Fackler,Jr.,Inorg.Chem.,1978, 17, 2182.

21.G.N. George, J. Byrd, D.R. Winge,J. Biol. Chem., 1988,263, 8199.

22.K.H. Nakagawa, C. Inouye, B. Hedman, M. Karin, T.D. Tullins, K.O. Hodgson, J.Amer.Chem.Soc.,1991, 113, 3621.

23.M. Kanatzidis, private information.

24.H. Schmidbaur, Gold Bulletin, 1990,23, 11.

25.R. Uson, A. Laguna, Coord. Chem. Rev., 1986, 70, 1.

Recent Developments in the Cluster Chemistry of Gold

D. Michael P. Mingos

DEPARTMENT OF CHEMISTRY, IMPERIAL COLLEGE OF SCIENCE
TECHNOLOGY AND MEDICINE, SOUTH KENSINGTON,
LONDON SW7 2AY, UK

1.   INTRODUCTION

Gold forms a wide range of cluster compounds when the metal has an oxidation state intermediate between 0 and +1. Athough in many compounds where gold has a formal oxidation state of +1 and a metal-metal bond does not have to be invoked in order to satisfy the requirements of the effective atomic number rule metal-metal contacts close to that found in the bulk metal (2.884Å) are observed. These metal-metal interactions between adjacent metals with d^{10} configurations are formally between atoms with completed electron shells and therefore should be repulsive, but of course gold has available empty 6s and 6p orbitals which can intervene to mitigate these antibonding interactions. The relativistic effects associated with these orbitals and in particular the contraction of the 6s valence orbitals makes an important contribution to the strength of the metal-metal interactions in these situations. In the gold cluster compounds where the formal oxidation state is less than +1 then the direct interactions between the 6s orbitals on the metal atoms are most important in deciding the closed shell requirements and structures of the clusters.

The cluster compounds of gold are generally stabilised by tertiary phosphine ligands and the first examples were reported by Malatesta and his coworkers in Milan in the late 1960's.[1] Since then many examples of gold cluster compounds have been characterised and a sufficient number are known for reliable bonding principles to have emerged. In addition during the last ten years many examples of high nuclearity heterometallic gold cluster compounds have been discovered. The geometries of these clusters can also be rationalised within the theoretical framework originally proposed for homo-metallic gold clusters, although interesting problems have arisen regarding the site preferences of the non-gold atoms within the cluster and require theoretical interpretation.

The synthesis and characterisation of gold cluster compounds with main group interstitial atoms has received considerable attention recently. Schmidbaur[2] has shown that the normal valence considerations associated with the main group atoms are violated and interesting structures result.

In this paper the structural and bonding aspects of homo- and heterometallic gold clusters with 4 to 13 metal atoms will be reviewed. Higher nuclearity clusters are discussed by other participants in the conference.

2.   HOMONUCLEAR CLUSTER COMPOUNDS OF GOLD

Cluster compounds of gold with 4 to 13 metal atoms are well established and those with more than seven metal atoms have an interstitial gold atom located in the centre of the cluster. The simplest clusters have a tetrahedron of metal atoms and bridging ligands on two of the edges, e.g. $[Au_4I_2(PPh_3)_4]$ and $[Au_4(SnCl_3)_2(PPh_3)_4]$. The latter which was recently characterised in our laboratories provides an unusual example of an $SnCl_3$ ligand acting as a bridge between two metal atoms.[3] In contrast, in the higher nuclearity cluster $[Au_8(SnCl_3)(PPh_3)_7]^+$ the $SnCl_3$ ligand bonds only to the interstitial gold atom.[4]

In the tetrahedral clusters the two valence electrons occupy a molecular orbital which results from the in phase overlap of 6s orbitals from the four gold atoms $[S^\sigma]^2$ and they are therefore formally derivatives of $[Au_4(PPh_3)_4]^{2+}$. For clusters with six and seven metal atoms, the presence of a single multicentred two electron bond is insufficient to stabilise the cluster geometry and additional electrons are required to occupy delocalised molecular orbitals which have a single node and therefore resemble p atomic wave functions. These molecular orbitals are therefore designated by the symbol P^σ.[5,6] A spherical cluster geometry therefore will only result when the P^σ set of three molecular orbitals is either completely empty or completely filled and partial occupation of the P^σ set results in non-spherical geometries. For example $[Au_6(PPh_3)_6]^{2+}$ has a prolate skeletal geometry based on two tetrahedra sharing an edge because the delocalised molecular orbitals which are occupied correspond to $[S^\sigma]^2[P^\sigma]^2$ and $[Au_7(PPh_3)_7]^+$ has an oblate geometry based on a pentagonal bipyramid corresponding to the electron configuration $[S^\sigma]^2[P^\sigma]^4$. For higher nuclearity clusters the relatively weak metal-metal bonding resulting from the overlap of 6s valence orbitals is enhanced by the introduction of an interstitial gold atom which strengthens the radial metal-metal bonds.[7] Naturally, this type of stabilisation is only available to those higher nuclearity clusters where the size of the interstitial hole is sufficiently large to accommodate a gold atom. Schmidbaur's compounds have interstitial main group atoms where the requirements for the cavity size are not so restrictive. In all of his compounds the delocalised molecular orbitals have the pseudo- symmetry labels $[S^\sigma]^2[P^\sigma]^6$ and therefore adopt spherical geometries based on the tetrahedron, trigonal bipyramid, square pyramid and octahedron.[8]

Higher nuclearity gold cluster compounds with the interstitial atom have geometries which may be classified as spherical or toroidal[9] and which also reflect the number of valence electrons occupying the molecular orbitals derived primarily from the 6s valence orbitals. The spherical clusters have a complete shell resulting from the electron configuration $[S^\sigma]^2[P^\sigma]^6$ and the toroidal clusters have the incomplete shell configuration $[S^\sigma]^2[P^\sigma]^4$ which leads to an oblate electron distribution which is compatible with a toroidal geometry. The dominance of the radial bonding in these clusters results in a soft potential energy surface for the interconversion of alternative polytopal geometries. The broad topological classifications of spherical or toroidal are therefore more appropriate than polyhedral classifications based on specific geometries when the energy differences separating different structures are very small. The structures of gold clusters based on the spherical-toroidal partitioning are summarised in Figure 1. It is significant that all of the structures may be related either to a centred chair or a centred crown of atoms. The spherical clusters have additional atoms located along the principal symmetry axes of these fragments and the toroidal clusters have the additional

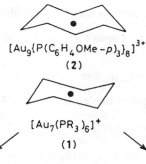

$[Au_9\{P(C_6H_4OMe-p)_3\}_8]^{3+}$
(2)

$[Au_7(PR_3)_6]^+$
(1)

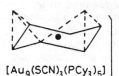

$[Au_8(PPh_3)_7]^{2+}$, p.e.c. 100

$[Au_8(PPh_3)_8]^{2+}$, p.e.c. 102

$[Au_9(SCN)_3(PCy_3)_5]$

$[Au_9(PPh_3)_8]^+$, p.e.c. 114

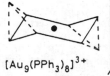

$[Au_9(PPh_3)_8]^{3+}$

p.e.c. 112

$[Au_{11}I_3(PPh_3)_7]$, p.e.c. 138

$[Au_{10}Cl_3(PCy_2Ph)_6]^+$, p.e.c. 124

$[Au_{13}Cl_2(PMe_2Ph)_{10}]^{3+}$, p.e.c. 162

Figure 1.   Classification of the structures of gold cluster compounds. The left hand column gives the toroidal (oblate) structures and the right hand column the spherical structures (p.e.c. represents the polyhedral electron count).

atoms located away from these axes in order to maintain the oblate nature of the structure. The structural flexibility associated with gold clusters has many consequences for the observed structures of these clusters in solution and the solid state.

Stereochemical non-rigidity of gold clusters

The stereochemical non-rigidity of gold clusters has several important manifestations. Firstly, alternative structures are observed for the clusters in the solid state depending on the precise nature of the ligands, the counter anions and the mode of crystallisation. For example, although the cluster $[Au_{11}I_3(P\{p\text{-}C_6H_4F\}_3)_7]$ has a C_{3v} geometry in the solid state[10] which may be described in terms of an icosahedron with one triangular face replaced by a single gold atom, the related gold cluster $[Au_{11}(PMePh_2)_{10}]^{3+}$ [11] has a more conventional deltahedral D_{4d} geometry based on a square antiprism. Although the latter has two distinct phosphorus chemical environments in the ratio 2:8 the $^{31}P\{^1H\}$ n.m.r. spectrum shows only a single resonance suggesting either that these environments have very similar chemical shifts or the compound is stereochemically non-rigid even in the solid state.[12] An examination of the detailed cluster geometries in these C_{3v} and D_{4d} variants shows that only small atomic motions are required to interconvert them. Furthermore, the process is symmetry allowed because both geometries are associated with $[S^\sigma]^2[P^\sigma]^6$ closed shell configurations.[13] Both of these factors contribute to the low activation energy for the polytopal rearrangement. In solution the compounds also show only a single phosphorus resonance suggesting that they are also stereochemically non-rigid in solution.

The structures of the toroidal gold clusters $[Au_9(PAr_3)_8]^{3+}$ depend on the counter ion and either a D_{2h} geometry derived from the removal of 4 vertices from an icosahedron or an alternative crown geometry derived from removal of the two low connectivity vertices of the square antiprism observed for $[Au_{11}(PMe_2Ph)_{10}]^{3+}$ are observed. Indeed with the correct combination of counterion and phosphine it is possible to isolate both skeletal isomers for the same compound. The compound $[Au_9(P\{p\text{-}C_6H_4OMe\}_3)_8](NO_3)_3$ thereby provided the first example of skeletal isomerism in cluster chemistry.[14] The two isomers do not have identical crystallographic cell volumes and therefore it has proved possible to interconvert them in the solid state using large hydrostatic pressures.[15] The transformation may be conveniently monitored using differences in the electronic spectra of the two isomers. More recently we have reported the interconversion of a similar pair of isomeric rhodium-gold cluster compounds by exerting high pressures on solid state samples.[16]

The only gold clusters which are stereochemically rigid at room temperature and on the n.m.r. time scale are those based on the icosahedron. The first icosahedral gold cluster was predicted on the basis of theoretical calculations in 1976 [7] and experimentally realised for $[Au_{13}Cl_2(PMe_2Ph)_{10}]^{3+}$ in 1981.[17] This compound was synthesised from $[Au_{11}(PMe_2Ph)_{10}]^{3+}$ and either $Au(PMe_2Ph)Cl$ or NEt_4Cl. If the $^{31}P\{^1H\}$ n.m.r. spectrum of $[Au_{13}Cl_2(PMe_2Ph)_{10}]^{3+}$ is recorded at low temperatures then the spectrum is consistent with the observed solid state structure based on an icosahedron with the *para-* gold vertices coordinated to chloride ligands. However, it is apparent from $^{31}P\{^1H\}$ n.m.r. studies that at room temperature the icosahedron disproportionates according to the following equation:

$$3[Au_{13}Cl_2(PMe_2Ph)_{10}]^{3+} \longrightarrow [Au_{11}(PMe_2Ph)_{10}]^{3+} + 2Au(PMe_2Ph)^+ + 2[Au_{13}Cl_3(PMe_2Ph)_9]^{2+}$$

Subsequently the corresponding icosahedral clusters $[Au_{13}Br_3(PMePh_2)_9]^{2+}$ and $[Au_{13}X_4(PMePh_2)_8]^+$ (X = Cl, Br or I) have been isolated and characterised by n.m.r. and FAB mass spectrometry.[18] In each case the n.m.r. evidence is consistent with a stereochemically rigid icosahedral structure. Interestingly the magnitude of the $^4J(P\text{-}P)$ coupling constants are most helpful in distinguishing alternative isomeric structures because their magnitudes depend on the angle subtended at the central gold atoms by the Au-P bonds which are involved in the spin-spin coupling. Furthermore, these compounds show no tendency to revert to the lower nuclearity $[Au_{11}(PMePh_2)_{10}]^{3+}$ cluster. The nuclearity of the gold clusters is very sensitive to the steric requirements of the ligands as defined by the "cluster cone angle" and presumably the replacement of phosphines by the less sterically demanding halides stabilises the icosahedral clusters.[19]

The importance of the steric requirements of the ligands in controlling the aggregation and fragmentation reactions of these gold clusters is underlined by the recent observation that the addition of $S_2C_2(CN)_2^{2-}$ to $[Au_9(PPh_3)_8]^{3+}$ does not lead to the elimination of a gold atom resulting from the formation of $[Au\{S_2C_2(CN)_2\}_2]^{3-}$ but the formation of $[Au_{10}\{S_2C_2(CN)_2\}_2(PPh_3)_7]$ with toroidal skeletal geometry based on a C_{3v} centred chair with three additional gold atoms bridging alternant edges.[20]

3. ICOSAHEDRAL GOLD-SILVER CLUSTER COMPOUNDS

The principles developed to understand the aggregation reactions of gold clusters have been recently utilised in the synthesis of the first icosahedral gold-silver cluster compounds. When $[Au_{11}(PMePh_2)_{10}]^{3+}$ is treated with 4 mol. equivalents of $[Ag(PMePh_2)Cl]$ in CH_2Cl_2 $[Au_9Ag_4Cl_4(PMePh_2)_8]^+$ is formed in high yield and isolated as the $C_2B_9H_{12}^-$ salt.[21] Cluster growth is promoted by the relative lability of the silver phosphine complex and the smaller cone angle of the residual Ag-Cl fragment relative to $Au(PMePh_2)$. A single crystal structural analysis and FAB mass spectrometry have confirmed the icosahedral geometry illustrated in Figure 2. The silver atoms are not disordered and the interstitial site is occupied by a gold atom. This ordered structure is consistent with the site preference arguments we have developed using elementary theoretical concepts.[22] The cluster core has C_{2v} symmetry with the silver atoms occupying the 1,2,8 and 10 positions. The radial metal-metal bond lengths are shorter than the tangential bonds which is a general characteristic of centred gold clusters. The $^{31}P\{^1H\}$ n.m.r. solution spectrum is consistent with the structure observed in the solid state and suggests that the cluster is stereochemically rigid on the n.m.r. time scale. Multiplets observed in the n.m.r. spectrum due to $^3J(Ag\text{-}P)$ couplings may be interpreted in terms of the angles subtended at the interstitial gold atom by the Ag-Cl and Au-P fragments. If the Au and Ag atoms have a *para-* relationship then $^3J(Ag\text{-}P)$ is of the order of 70Hz , but this drops to 17Hz when Au and Ag have a *meta-* relationship. This compound is structurally the parent of the high nuclearity gold-silver clusters based on linear, triangular and tetrahedrally linked icosahedra characterised by Teo and his coworkers.[23] The closed shell requirements of these condensed icosahedra have been estimated using molecular orbital theory arguments.[24]

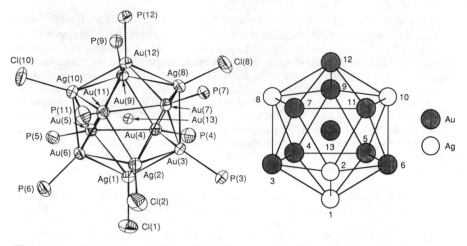

Figure 2   Structure of icosahedral $[Au_9Ag_4Cl_4(PMePh_2)_8]^+$

4. GOLD RHODIUM CLUSTERS DERIVED FROM THE ICOSAHEDRON

The co-reduction of $[AuCl(PPh_3)]$ and $[RhCl(CNC_8H_9)_3]$ leads to a mixture of rhodium-gold cluster compounds from which $[Rh(CNC_8H_9)_3(AuPPh_3)_5]^{2+}$ and $[Rh(CNC_8H_9)_2(AuPPh_3)_6(AuCl)_2]^+$ were isolated[25,26]. The related cluster $[Rh(CO)_2(AuPPh_3)_7]^{2+}$ was synthesised by the addition of of 1 equivalent of $[Au(PPh_3)(NO_3)]$ to the thf soluble fraction of a solution obtained by the borohydride reduction of a mixture of $[Au(PPh_3)(NO_3)]$ and $[Rh(CO)_2(CH_3CN)_2]^+$. The structures of the compounds are shown in Figure 3 and may be structurally related to the icosahedron as shown in Figure 4. $[Rh(CO)_2(AuPPh_3)_6]^+$ was not structurally determined and its structure has been inferred from spectroscopic and mass spectral data.

It is noteworthy that in this series of clusters and in contrast to the gold-silver clusters it is now the rhodium which occupies the interstitial site. This can be attributed to the stronger metal-metal bonds formed by rhodium and fits in with the site preference ideas developed theoretically. Although these clusters have the appropriate electron configuration to adopt spherical geometries, i.e. they have $[S^\sigma]^2[P^\sigma]^6$ ground state electronic configurations, they actually adopt hemispherical geometries with the gold atoms clustered around one half of the sphere. This occurs to enable the rhodium atom to remain co-ordinated to either the iso-cyanide or carbonyl ligands. This class of compound therefore provides a nice illustration of the way in which the site preference effects in clusters are not only determined by the relative strengths of the metal-metal bonds, but also by ligand-metal binding effects. Both isocyanide and carbon monoxide bond more strongly to rhodium than gold and therefore maximum cluster stability may be achieved when the rhodium binds to as many gold atoms as possible and also maintains its co-ordination to the π-acid ligands. The hemispherical geometry provides the means of satisfying these two requirements. $^{31}P\{^1H\}$ n.m.r. experiments have indicated that these compounds are stereochemically non-rigid in solution , although in some cases it has proved possible to freeze out the fluxional process at low temperatures.

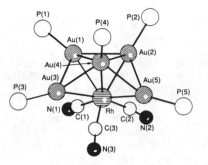

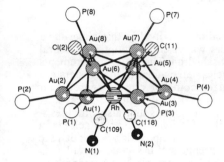

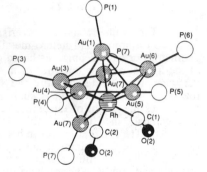

Figure 3   Structures of gold-rhodium clusters

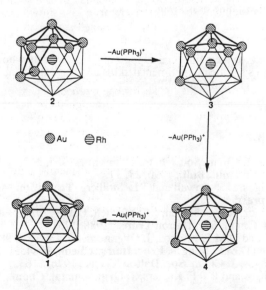

Figure 4.   Relationship of the structures of the gold-rhodium clusters to the icosahedron.

4.   ICOSAHEDRALLY   BASED   GOLD-PALLADIUM   CLUSTERS

The icosahedral theme may be completed by discussing some recent results in the area of gold palladium cluster compounds. The addition of $[AuPPh_3]^+$ to $[Pd_8(CO)_8(PMe_3)_7]$ has resulted in the isolation of the novel cluster $[Pd_{14}Au_2(PMe_3)_{11}(\mu-CO)_2(\mu_3-CO)_7]^{2+}$ which is based on a $Pd_{11}Au_2$ icosahedron with an additional palladium triangle fused to it through a face. It is most unusual for cluster compounds of palladium to adopt icosahedral geometries. This cluster is particularly interesting because the gold atoms are not bonded to any ligands and are stabilised solely by metal-metal bonds.[27]

5.   SUMMARY

The synthetic and characterisational studies described in this paper have focussed on homo- and hetero-metallic clusters which either adopt icosahedral geometries or have geometries based on the icosahedron. The steric requirements of the ligands and the labilities of the mononuclear precursors to the clusters are very important in determining the successful outcome of cluster aggregation reactions. The icosahedron is such a ubiquitous structural template because it maximises the extent of tangential metal-metal interactions and also provides a very compact shell which enables the radial metal-metal bonding to achieve a maximum. The site preferences in heteronuclear cluster compounds suggest that the most important determining factor is the relative strengths of the metal-metal bonds formed by the constituent metal atoms. Superimposed on this is the strength of metal-ligand bonds. In particular a metal atom which on the basis of metal--metal bonding may prefer an interstitial site may be encouraged to migrate to a hemispherical surface site if it can form strong bonds preferentially to one of the ligands. The rhodium-gold clusters described above have provided examples of this behaviour. Finally the study of palladium-gold cluster compounds has revealed an interesting example of gold atoms occupying surface sites on the cluster, whilst not being stabilised by phosphine or isocyanide ligands.

Acknowledgements

The SERC is thanked for their financial support and my students whose names are given in the references for their experimental skills and dedication and in particular Roy Copley, Jane Haggitt and Chris Hill.

REFERENCES

1.   L. Malatesta, **J. Chem. Soc. Chem. Commun.**, 1965, 212.
2.   H. Schmidbaur, **Gold. Bull.**, 1990, *23*, 11.
3.   D.M.P. Mingos, H.R. Powell and T.L. Stolberg, **Transition Met. Chem.**, *i n press*.
4.   Z. Demidowicz, R.L. Johnston, J.C. Machell, D.M.P. Mingos and I.D. Williams, **J. Chem. Soc. Dalton Trans.**, 1988, 1751.
5.   D.G. Evans and D.M.P. Mingos, **J. Organometal. Chem.**, 1985, *295*, 389.
6.   K.P. Hall and D.M.P. Mingos, **Prog. Inorg. Chem.**, 1984, *32*, 237.
7.   D.M.P. Mingos, **J. Chem. Soc. Dalton Trans.**, 1976, 1163.
8.   D.M.P. Mingos and R.P.F. Kanters, **J. Organometal. Chem.**, 1990, *338*, 405.
9.   C.E. Briant, K.P. Hall, D.M.P. Mingos and A.C. Wheeler, **J. Chem. Soc. Chem. Commun.**, 1984, 248.

10. M. McPartlin, R.Mason,and L. Malatesta, **J. Chem. Soc. Chem. Commun.**, 1969, 334.
11. D.M.P. Mingos and R.C.B. Copley, *unpublished results*.
12. N.J. Clayden, C.M. Dobson, K.P. Hall, D.M.P. Mingos and D.J. Smith, **J. Chem. Soc. Dalton Trans.**, 1985, 1811.
13. D.J. Wales, D.M.P. Mingos and Lin Zhenyang, **Inorg. Chem.**, 1989, *28*, 2748.
14. C.E. Briant, K.P. Hall and D.M.P. Mingos, **J. Chem. Soc. Chem. Commun.**, 1984, 290.
15. J.L. Coffer, H.G. Drickamer and J.R. Shapley, **Inorg. Chem.** , 1990, *29*, 3900.
16. K.L. Bray, H.G. Drickamer, D.M.P. Mingos M.J. Watson and J.R. Shapley, **Inorg. Chem.**, 1991, *30*, 864.
17. C.E. Briant, B.R.C. Theobald, J.W. White, L.K. Bell and A.J. Welch, **J. Chem. Soc. Chem. Commun.**, 1981, 201.
18. R.C.B. Copley and D.M.P. Mingos, *unpublished results*.
19. D.M.P. Mingos, **Inorg. Chem.**, 1982, *21*, 466.
20. J.L. Haggitt and D.M.P. Mingos, *unpublished results*.
21. R.C.B. Copley and D.M.P. Mingos, **J. Chem. Soc. Dalton Trans.**, 1992, 1755.
22. D.M.P. Mingos and Lin Zhenyang, **Comments in Inorganic Chemistry**, 1989, *34*, 72.
23. B.K. Teo, M.C. Hong, H. Zhang,and D.B. Huang, **Angew. Chem. Int. Ed., Engl.** 1987, *26*, 897.
24. Lin Zhenyang, R.P.F. Kanters and D.M.P. Mingos, **Inorg. Chem.**, 1991, *30*, 91.
25. S.G. Bott, D.M.P. Mingos and M.J. Watson, **J. Chem. Soc. Chem. Commun.**, 1989, 1192.
26. S.G. Bott, H. Fleischer, M. Leach, D.M.P. Mingos, H.R. Powell, D.J. Watkin and M.J. Watson, **J. Chem. Soc. Dalton Trans.**, 1991, 2569.
27. C.M. Hill and D.M.P. Mingos, *unpublished results*.

Gold Oxo, Imido, and Hydrazido Complexes and Gold Clusters

P. R. Sharp, Y. Yi, Z. Wu, and V. Ramamoorthy

DEPARTMENT OF CHEMISTRY, UNIVERSITY OF MISSOURI –
COLUMBIA, COLUMBIA, MO 65211, USA

1 INTRODUCTION

As a synthetic group, we are interested in developing homogeneous models for the heterogeneous surface chemistry of late-transition-metals. Of particular importance is the ability of many late-transition-metals to adsorb and dissociatively activate molecular oxygen (Scheme 1). Oxo and the iso-

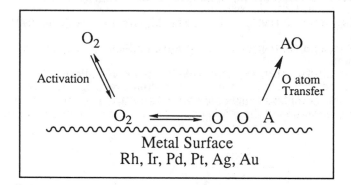

Scheme 1

electronic imido complexes are potential models for the surface oxygen atoms. Consequently, we have been studying the synthesis and reactivity of oxo and imido complexes of several of these catalytically important metals. In addition to Rh[1] and Pt[2], we are interested in Au complexes and our attention has naturally turned to the one Au oxo complex[3] to be found in the literature, $[\{(Ph_3P)Au\}_3(\mu\text{-}O)]^+$. Our investigations into the chemistry of this complex have led to the synthesis of a number of new complexes and the discovery of simple, high yield methods for the preparation of several gold cluster complexes. Among the new complexes are the first examples of isolated intermediates in the reduction of Au(I) to gold clusters.

2 RESULTS

Oxo, Imido and Hydrazido Complexes

There are several preparative routes to $[\{(Ph_3P)Au\}_3(\mu\text{-}O)]^+$ and we find that the route shown in eq 1 readily accommodates the use of other phosphine ligands. An exception is for L = (o-MePh)$_3$P where the oxo complex is prepared in a 2 step procedure from AgBF$_4$ and NaOH.

$$LAuCl \xrightarrow{\quad Ag_2O,\ BF_4^- \quad} [(LAu)_3(\mu\text{-O})]BF_4 \qquad (1)$$

$$L = PPh_3,\ PPh_2Me,\ PPhMe_2,\ PPh_2Et,\ (p\text{-ClPh})_3P$$

New oxo complexes are also formed in mixtures by exchange between L'AuCl and [(LAu)$_3$(μ-O)]$^+$ (eq 2) and this provides a convenient check on the

$$3L'AuCl + [(LAu)_3(\mu\text{-O})]^+ \Longleftrightarrow 3LAuCl + [(L'Au)_3(\mu\text{-O})]^+ \quad (2)$$

$$L = PPh_3 \qquad L' = PPh_2Me,\ PPhMe_2,$$

stability of target oxo complexes. Although four oxo complexes, corresponding to all possible combinations of LAu and L'Au, should be present in the mixtures only two peaks are observed in the ^{31}P NMR spectra indicating rapid exchange or peak coincidence.

The oxo complexes may be converted into the analogous imido complexes by the procedures in eq 3 and 4.[4] The first procedure was independently discovered by two other groups for L = PPh$_3$.[5,6]

$$[(LAu)_3(\mu\text{-O})]^+ + RNH_2 \longrightarrow [(LAu)_3(\mu\text{-NR})]^+ + H_2O \qquad (3)$$

$$[(LAu)_3(\mu\text{-O})]^+ + RNCO \longrightarrow [(LAu)_3(\mu\text{-NR})]^+ + CO_2 \qquad (4)$$

$$L = PPh_3,\ PPh_2Me,\ PPhMe_2 \qquad R = alkyl\ or\ aryl$$

Hydrazido complexes are obtained in reactions (Scheme 2) similar to those for the preparation of the imido complexes. All three types have been

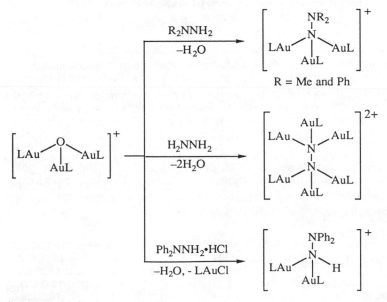

Scheme 2

structurally characterized by X-ray diffraction for L = PPh$_3$ and R = Ph.

Gold Clusters

Both the oxo and the imido complexes (L = PPh$_3$) react with CO at 3 atm to give the red cluster complex, [L$_6$Au$_6$]$^{2+}$ in >90% yield (Scheme 3) representing a considerable improvement over previous syntheses.[7]

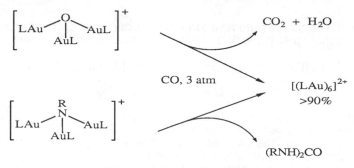

$$CO_2 + H_2O$$

CO, 3 atm

$$[(LAu)_6]^{2+}$$
>90%

$$(RNH)_2CO$$

L = PPh$_3$,   R = Ph

Scheme 3

The hydrazido complexes give gold clusters by decomposition. The dimethyl hydrazido complexes decompose cleanly (THF) to [(LAu)$_6$]$^{2+}$, [(LAu)$_{10}$Au]$^{3+}$, or both, again in yields >90% (Scheme 4). Although preliminary experiments

$$[(LAu)_3NNMe_2]^+ \xrightarrow{-Me_2NN=NNMe_2}$$

[(LAu)$_6$]$^+$
L = PPh$_3$, (p-ClPh)$_3$P, PPh$_2$Et

[(LAu)$_{10}$Au]$^{3+}$ + [L$_2$Au]$^+$
L = PPh$_2$Me, PPhMe$_2$, PPh$_2$Et

Scheme 4

indicated second order decomposition,[8] more recent results show first order kinetics. The nature of the cluster produced in the decompositions is phosphine cone angle dependent with the smaller cone angle phosphines giving the [(LAu)$_{10}$Au]$^{3+}$ clusters, larger cone angle phosphines giving the [(LAu)$_6$]$^{2+}$ clusters and intermediate cone angles giving mixtures of both clusters.

The interaction of the more sterically crowded hydrazine, (i-Pr)$_2$NNH$_2$, with the gold oxo complexes does not produce detectable quantities of the expected hydrazido complexes. Instead, direct production of the same gold clusters obtained from the decomposition of the dimethyl hydrazido complexes is observed.

3 DISCUSSION

Our discovery of the clean decomposition of the dialkyl hydrazido complexes and our synthesis of a variety of gold oxo precursors offers an excellent opportunity to investigate the formation pathways of gold clusters. The first order decomposition kinetics of [(LAu)$_3$(μ-NNMe$_2$)]$^+$ supports earlier thoughts that gold clusters form by the aggregation of small, reduced fragments.[9,10,11] However, there still remain questions about the nature of these fragments, how they come together and the factors that determine which cluster or clusters are produced.

Our results confirm earlier indications that cone angle is a major determining factor.[12] The larger cone angle phosphines favor the formation of the smaller $[(LAu)_6]^{2+}$ clusters and the smaller cone angle phosphines favor the formation of the $[(LAu)_{10}Au]^{3+}$ clusters. The result with the intermediate phosphine PPh_2Et, the formation of both clusters, is significant. We had thought that the $[(LAu)_{10}Au]^{3+}$ clusters formed by the dimerization of the $[(LAu)_6]^{2+}$ clusters followed by elimination of L_2Au^+. This is clearly incorrect as both clusters are stable and they do not readily interconvert. Another possibility, that the $[(LAu)_{10}Au]^{3+}$ clusters are formed from aggregation of the $[(LAu)_6]^{2+}$ clusters with smaller fragments, is possible provided that this is a fast reaction in comparison to the formation of the $[(LAu)_6]^{2+}$ cluster. This is required since both clusters seem to form simultaneously for L = PPh_2Et. Scheme 5 is based on this pathway. Alternatively, the two clusters could form

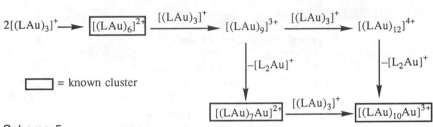

Scheme 5

from a common precursor along competing pathways. No matter what the pathway to these clusters, the high selectivity of the reactions is remarkable.

REFERENCES

1. Leading ref.: Y.-W. Ge, P. R. Sharp, Inorg. Chem., 1992, 31, 379.
2. W. Li, C. L. Barnes, P.R. Sharp, J. Chem. Soc., Chem. Commun., 1990, 1634.
3. A. N. Nesmeyanov, E. G. Perevalova, Yu. T. Struchkov, M. Yu. Antipin, K. I. Grandberg, V. P. Dyadchenko, J. Organomet. Chem., 1980, 201, 343.
4. V. Ramamoorthy, P. Sharp, Inorg. Chem., 1990, 29, 3336.
5. A. Grohmann, J. Riede, H. Schmidbaur, J. Chem. Soc., Dalton Trans.,1991, 783.
6. E. G., Perevalova, K. I. Grandberg, E. I. Smyslova, D. N. Kuz'mina, Organomet. Chem. USSR, 1989, 29, 523.
7. C. E. Briant, K. P. Hall, D. M. P. Mingos and A. C. Wheeler, J. Chem. Soc., Dalton Trans.,1986, 687.
8. V. Ramamoorthy, Z. Wu, Y. Yi, P. R. Sharp, J. Am. Chem. Soc., 1992, 114, 1526.
9.. J.J. Steggerda, J.J. Bour, J.W.A. van der Velden, Recl.: Neth. Chem. Soc., 1982, 101, 164.
10. D. M. P. Mingos, Polyhedron, 1984, 12, 1289.
11. K. P. Hall, D. M. P. Mingos, Prog. Inorg. Chem., 1984, 32, 237.
12. D. M. P. Mingos, Inorg. Chem., 1982, 21, 464.

Mixed-metal Clusters Containing Osmium and Gold

Angelo J. Amoroso[1], Jack Lewis[1], Paul R. Raithby[1], and Wing-Tak Wong[2]

[1] UNIVERSITY CHEMICAL LABORATORY, LENSFIELD ROAD, CAMBRIDGE CB2 1EW, UK
[2] DEPARTMENT OF CHEMISTRY, THE UNIVERSITY OF HONG KONG, POKFULAM ROAD, HONG KONG

INTRODUCTION

The reaction of group IB metal fragments with anionic metal carbonyl clusters to give heterometallic metal clusters is well known.[1]   However, the bimetallic cationic fragment containing a bidentate phosphine such as $[Au_2(dppe)]^{2+}$, dppe=$Ph_2PCH_2CH_2PPh_2$, has only been used recently in a few cases with ruthenium and osmium carbonyl clusters in which both Au atoms are bonded to the same cluster anion. [2-4] We would like to investigate the possibility of linking some osmium carbonyl cluster anions by these cations as a method of generating high nuclearity heterometallic species. Here we describe the preparations and full characterisation of

(i) $[\{H_3Os_4(CO)_{12}\}_2Au_2(dppe)]$ **1** dppe=$Ph_2PCH_2CH_2PPh_2$

(ii) $[\{HOs_4(CO)_{12}Au(dppa)\}_2]$ **2** dppa=$Ph_2P-C\equiv C-PPh_2$

(iii) $[Os_{10}C(CO)_{24}Au_2(dppm)]$ **3** dppm=$Ph_2PCH_2PPh_2$

RESULTS AND DISCUSSION

Linked Cluster: $[\{H_3Os_4(CO)_{12}\}_2Au_2(dppe)]$ **1**

Reaction of $[N(PPh_3)_2][H_3Os_4(CO)_{12}]$ with $[Au_2(dppe)Cl_2]$ in a 2:1 ratio in the presence of excess $TlPF_6$ (halide acceptor) gives the air-stable yellow product **1** in almost quantitative yield with respect to the Au reagent used, after chromatography on silica. A single crystal X-ray analysis of **1** shows that two tetra-osmium units are linked together by a $[Au_2(dppe)]^{2+}$ fragment. The molecular structure of **1** is shown in Figure 1, together with some selected bond parameters. The structural features of $[H_3Os_4(CO)_{12}AuPR_3]$ unit in **1** are essentially the same as in $[H_3Os_4(CO)_{12}AuPEt_3]$.[5] The spectroscopic data of **1**

are fully consistent with the solid state structure, see
Table 1.

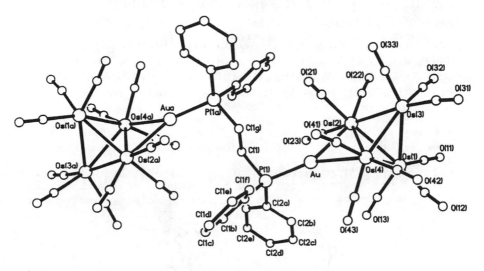

Figure 1
The molecular structure of [{H₃Os₄(CO)₁₂}₂Au₂(dppe)] **1**.

Selected Bond Parameters

Bond lengths (Å)
Au–Os(2)	2.778(2)	Au–Os(4)	2.789(2)
Os(1)–Os(2)	2.961(1)	Os(1)–Os(3)	2.956(1)
Os(1)–Os(4)	2.812(2)	Os(2)–Os(3)	2.811(2)
Os(2)–Os(4)	2.942(1)	Os(3)–Os(4)	2.960(2)
Au–P(1)	2.288(6)	P(1)–C(1)	1.85(2)
C(1)–C(1g)	1.56(3)		

Bond angles (°)
Os(2)–Au–Os(4)	63.8(1)	Os(2)–Au–P(1)	143.4(1)
Os(4)–Au–P(1)	152.8(1)		

Linked Cluster in Cyclic Fashion:[{HOs₄(CO)₁₂Au₂(dppa)}₂] **2**

Reaction of [N(PPh₃)₂][H₃Os₄(CO)₁₂] with
[Au₂(dppa)Cl₂], in the presence of excess Et₃N (for
deprotonation) and TlPF₆ (halide acceptor), in CHCl₃ under
reflux gives the red air-stable compound **2** as the only
isolated product (*ca.* 30%). The spectroscopic data of **2** are
summarised in Table 1. The molecular structure of **2** is
shown in Figure 2, together with some important bond
parameters. The structure consists of two tetrahedral Os₄
units linked by two "Audppa" moieties. The gold atom of the
Audppa group bridges one edge of the Os₄ cluster unit while
the phosphorus atom at the other end of the Audppa unit

coordinates to one of the Os atoms of another Os_4 cluster to form a linkage between two Os_4 units. The same linkage is present at the other end of the two Os_4 clusters so that the compound **2** has a cyclic arrangement. There are two asymmetric bridging CO ligands in each Os_4 unit. It is worth noting that the reaction involves the loss of two H atoms and the addition of a phosphine, so that **2** may be considered as a substituted derivative of the $[Os_4H(CO)_{13}]^-$ anion. The $^{31}P\{^1H\}$ NMR spectrum of **2** shows two singlets at -149.0 ppm and -84.6 ppm that were assigned to the phosphorus coordinated to the osmium atom and gold atom respectivey. The ^1H NMR of **2** in CD_2Cl_2 shows only one singlet at -20.51ppm at room temperature and -50°C. However, the position of hydride atom cannot be determined reliably from X-ray analysis or potential energy calculations. The formation of **2** inevitably involves the loss of one Au atom from $Au_2dppaCl_2$. However, it has not been possible to establish the fate of this gold atom.

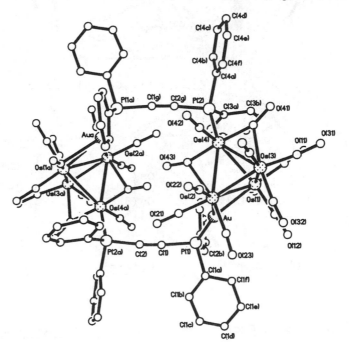

Figure 2
The molecular structure of $[\{HOs_4(CO)_{12}Au_2(dppa)\}_2]$ **2.**

Selected Bond Parameters

Bond lengths (Å)

Os(1)-Os(2)	2.933(3)	Os(1)-Os(3)	2.819(3)
Os(1)-Os(4)	2.976(3)	Os(2)-Os(3)	2.832(3)
Os(2)-Os(4)	2.834(3)	Os(3)-Os(4)	2.812(2)
Au-Os(1)	2.744(3)	Au-Os(2)	2.784(3)

Non-Linking Cluster:[Os$_{10}$C(CO)$_{24}$Au$_2$(dppm)] **3**.

Reaction of [N(PPh$_3$)$_2$][Os$_{10}$C(CO)$_{24}$] with 0.55 equivalent [Au$_2$(dppm)Cl$_2$] in the presence of excess TlPF$_6$ gives the red product **3** in approximately 25% yield after chromatographic separation on silica. A single crystal X-ray structure analysis revealed an Os$_{10}$ metal core bicapped with two Au atoms from the [Au$_2$(dppm)]$^{2+}$ fragment. The Au-Au distance is 2.971(1)Å. The molecular structure of **3** is shown in Figure 3, together with some important bond parameters. The spectroscopic data of **3** are fully consistent with the solid state structure, see Table 1.

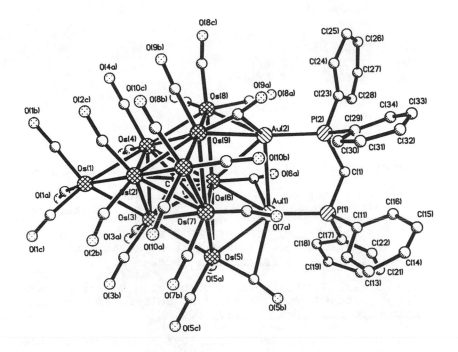

Figure 3
The molecular structure of [Os$_{10}$C(CO)$_{24}$Au$_2$(dppm)] **3**.

Selected Bond Parameters:

Bond lengths (Å)

Au(1)-Au(2)	2.971(1)	Au(1)-Os(5)	2.851(1)
Au(1)-Os(6)	2.863(1)	Au(1)-Os(7)	2.850(1)
Au(2)-Os(6)	3.030(1)	Au(2)-Os(8)	2.917(1)
Au(2)-Os(9)	2.751(1)		

Table 1   Spectroscopic data for **1,2** and **3**.

Compound	IR (υCO)[a]	MS (m/z)	^{1}H NMR[b]	^{31}P{^{1}H} NMR[c]
1	2094m,2071s 2033m,2004m 1970w,1950m,br	N/A	−20.27(s,6H) 2.81(m,4H) 7.49(m,20H)	−69.3(s)
2	2069s,2033vs 2020vs,1988m 1965w,1932m	3378	−20.51(s,2H) 7.83(m,40H)	−149.0(s,2P) −84.6(s,2P)
3	2091m,2068s 2053vs,2018vs 2008m	N/A	3.21(m,2H) 7.52(m,20H)	−81.3(s)

[a] solvent:CH_2Cl_2,cm^{-1}
[b] solvent:CD_2Cl_2,ref. $SiMe_4$
[c] solvent:CD_2Cl_2,ref. $P(OMe)_3$

REFERENCES

1. (a) C.E. Coffey, J. Lewis, R.S. Nyholm, J. Chem. Soc.,
        1964,1741.
    (b) B.F.G. Johnson, J. Lewis, W.J.H. Nelson, P.R.
        Raithby, M.D. Vargas, J. Chem. Soc., Chem. Comm.,
        1983,608.
    (c) B.F.G. Johnson, J. Lewis, W.J.H. Nelson, M.D.
        Vargas, D.Braga, K. Henrick, M. McPartlin, J
        Chem. Soc., Dalton Trans., 1986, 975.

2. P.A. Bates, S.S.D. Brown, A.J. Dent, M.B. Hursthouse,
    G.F.M. Kitchen, A.G. Orpen, I.D. Salter, V. Sik, J.
    Chem. Soc.,Chem.Comm., 1986, 600.

3. P.A. Bates, S.S.D. Brown, D.B. Dyson, M.B. Hursthouse,
    R.V. Parish, I.D. Salter, J. Chem. Soc., Dalton Trans.,
    1988, 1795.

4. S.S.D. Brown, S. Hudson, M. McPartlin, I.D. Salter, J.
    Chem.Soc., Dalton Trans., 1987,1967.

5. A. Cowie, B.F.G. Johnson, J.Lewis, P.R. Raithby,
    unpublished result.

ACKNOWLEDGEMENTS

    W.T.W. thanks the finanical support from The University
of Hong Kong.

Platinum–Gold Clusters Containing Copper and Silver

T. G. M. M. Kappen, M. F. J. Schoondergang,
P. P. J. Schlebos, J. J. Bour, and J. J. Steggerda

DEPARTMENT OF INORGANIC CHEMISTRY, UNIVERSITY OF
NIJMEGEN, TOERNOOIVELD 1, 6525 ED NIJMEGEN, THE NETHERLANDS

Introduction

The stability of clusters of the type $Pt\text{-}Au_n$, where Pt is in the centre surrounded by 6-9 Au atoms, is generally thought to be partly due to peripheral Au-Au interactions. These "$d^{10}\text{-}d^{10}$" interactions are considered to be attractive and important for Au-Au contacts. The question arose whether Ag and Cu can be substituted for the Au atoms while maintaining the stability of these cluster compounds.

Several of these Pt-Au-M (M=Cu, Ag) clusters have been synthesized and characterized. We will report about results of introducing Ag and Cu in the periphery of these clusters.

Introducing Ag and Cu into Pt-Au clusters

The reaction of $[Pt(AuPPh_3)_8]^{2+}$ with $AgNO_3$ or CuCl results in the mixed clusters $[Pt(AgNO_3)(AuPPh_3)_8]^{2+}$ and $[Pt(CuCl)(AuPPh_3)_8]^{2+}$ respectively.[1,2] Both have a 9-coordinated central Pt, and are stable in solution and air contact. The crystal structure of $[Pt(CuCl)(AuPPh_3)_8](NO_3)_2$ is given in Figure 1. The toroidal geometry, as calculated by the topological program TORUS,[3] is in accordance with the electron count of this cluster (16 electrons).

Both $[Pt(AgNO_3)(AuPPh_3)_8]^{2+}$ and $[Pt(CuCl)(AuPPh_3)_8]^{2+}$ react very fast with CO to yield the 18 electron clusters $[Pt(CO)(AgNO_3)(AuPPh_3)_8]^{2+}$ and $[Pt(CO)(CuCl)(AuPPh_3)_8]^{2+}$ respectively,[1,2] which have a spheroidal geometry, in accordance with their electron count. The solid state structure of $[Pt(CO)(CuCl)(AuPPh_3)_8](NO_3)_2$ is given in Figure 2.

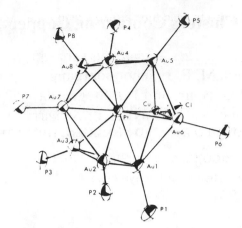

Figure 1. Crystal structure of $[Pt(CuCl)(AuPPh_3)_8](NO_3)_2$. Phenyl rings and NO_3-ions have been omitted for the sake of clarity. Thermal ellipsoids are at 50% probability.

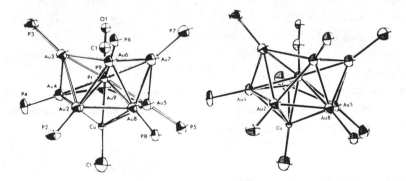

Figure 2. Crystal structure of $[Pt(CO)(CuCl)(AuPPh_3)_8](NO_3)_2$. Both figures are slightly rotated as compared to each other. Phenyl rings and NO_3-ions have been omitted for the sake of clarity. Thermal ellipsoids are at 50% probability.

The clusters $[Pt(CO)(AgNO_3)_2(AuPPh_3)_7]^+$ and $[Pt(CO)(CuCl)_2(AuPPh_3)_7]^+$ contain two Ag or Cu atoms among the 9 peripheral metal atoms. $[Pt(CO)(CuCl)_2(AuPPh_3)_7]^+$ can be prepared according to scheme 1. The same reaction scheme holds for $[Pt(CO)(AgNO_3)_2(AuPPh_3)_7]^+$, using $AgNO_3$ instead of CuCl. Both 18 electron $Pt(CO)M_2(AuP)_7$ clusters are stable in solution and air contact.

A related cluster, $[Pt(CO)(AuCl)_2(AuPPh_3)_7]^+$, was formed in different types of reaction as were used to yield the corresponding Ag and Cu compounds.[4]

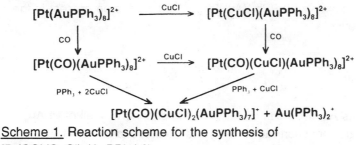

Scheme 1. Reaction scheme for the synthesis of $[Pt(CO)(CuCl)_2(AuPPh_3)_7]^+$.

Treatment of $[Pt(AuPPh_3)_8]^{2+}$ with three equivalents of $AgPPh_3NO_3$ in acetone leads to the formation of the 18 electron cluster $[Pt(PPh_3)(AgNO_3)_3(AuPPh_3)_6]$. The crystal structure, given in Figure 3, shows a 10-coordinated central Pt-atom and although there are no direct Ag-Ag contacts, there are 6 Au-Au contacts (2.81-2.90 Å) and 12 Au-Ag contacts (2.81-3.03 Å). In accordance with the electron count the geometry of this cluster can be classified as spheroidal.

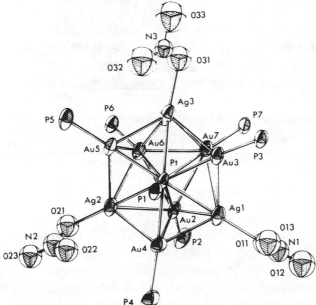

Figure 3. Crystal structure of $[Pt(PPh_3)(AgNO_3)_3(AuPPh_3)_6]$. Phenyl rings have been omitted for the sake of clarity. Thermal ellipsoids are at 50% probablity.

This cluster is stable in air and in the solvents benzene and toluene; when the cluster is dissolved in dichloromethane, silver metal is formed within a few hours and mainly $[Pt(PPh_3)(AuPPh_3)_6]^{2+}$ is found to be present in the resulting solution.

Conclusion

Up to 3 atoms Ag or Cu can be built into the peripheral shell of $Pt-Au_n$ clusters. Apart from short radial Pt-Ag and Pt-Cu bond lengths there are many (up to 12) short Au-Ag and Au-Cu contacts in the periphery. The compounds are stable and have toroidal or spheroidal geometries in accordance with their 16 or 18 cluster valence electrons. The Ag atoms are accessible due to the easily leaving NO_3-ions.

References

(1)   R.P.F. Kanters, P.P.J. Schlebos, J.J. Bour, W.P. Bosman, J.M.M. Smits, P.T. Beurskens and J.J. Steggerda, Inorg. Chem., 1990, 29, 324.

(2)   M.F.J. Schoondergang, J.J. Bour, P.P.J. Schlebos, A.W.P. Vermeer, W.P. Bosman, J.M.M. Smits, P.T. Beurskens and J.J. Steggerda, Inorg. Chem., 1991, 30, 4704.

(3)   R.P.F. Kanters and J.J. Steggerda, J. Cluster Science, 1990, 1, 229.

(4)   M.F.J. Schoondergang, PhD Thesis, University of Nijmegen, The Netherlands, 1992.

Clusters of Clusters: Coining Coinage Metal Clusters

Boon K. Teo*, Hong Zhang, and Xiaobo Shi

DEPARTMENT OF CHEMISTRY, UNIVERSITY OF ILLINOIS AT CHICAGO, CHICAGO, IL 60680, USA

1 INTRODUCTION

The last decade has seen tremendous progress in *cluster* research due to the fact that the collective behavior of a cluster often differs significantly from that of its constituents. In other words, the "whole" -- the cluster -- is often more than the "sum" of the "parts" -- the constituents -- due to synergic effects.

A cluster may be defined as an aggregate of atoms or molecules.[1] In general, atoms (or molecules) within a cluster can be held together by ionic, covalent, metallic, or hydrogen bonding, as well as weak van der Waals interactions. Recent advances in cluster research include gas-phase (bare) clusters,[2-9] ligated clusters,[10-18] encapsulated (e.g., in zeolites) clusters,[19-22] and extended (solid-state) clusters.[23-25] Some clusters are naked (for example, Bi_8^{2+} [26] and Ge_9^{2-} [27]) while others are ligated with ligands such as phosphines (e.g., $[Au_{13}Cl_{12}(PMe_2Ph)_{10}]^{3+}$ [28]) and carbonyls (e.g., $[Pt_{19}(CO)_{22}]^{4-}$ [29]). They can either be discrete clusters as in solution or extended clusters as in solids. Recent interests in carbon clusters, in particular, have produced many single-shell fullerenes[30] such as C_{60} and C_{70}, as well as multi-shell (nested) fullerenes which are either onionlike[31] or tubular[32] in shape.

Generally speaking, clusters represent the initial stages of fine particle formation.[33] Systematic studies of the preparation, structure, and properties of clusters will shed light on the mechanisms of *nucleation* and *growth* processes of fine particles[34-37] as well as factors which influence their size and morphology, their atomic arrangements and electronic structures, and their physical properties and chemical reactivities.[34-37]

Large metal clusters[38-53] (containing, for example, dozens of metal atoms) are of particular interest in that they lie in the region of aggregation between individual atoms and bulk metal. Investigations of the chemical and physical properties of these clusters will therefore provide new insights regarding how, where, and when metallic (or other bulk) behavior begins or ends.[54] Such basic understanding will ultimately allow one to modify the metallic behavior, to control the growth of metal particles, or to tailor make new metal or metal alloy phases. It is generally believed, and in many cases demonstrated, that unusual size-dependent properties can occur in the intermediate size range where there are tens to thousands of atoms (the so-called *quantum-size effect*).[55-60]

Recent developments in cluster synthesis have produced many high nuclearity metal clusters, some with metal arrangements resembling fragments of metallic lattices while others with metal frameworks distinctly different from that of the bulk.[38-53] A particular class of metal clusters is the heteronuclear or mixed-metal clusters containing more than one type of metals which may be referred to as *metal alloy clusters*. The formation of metal alloy clusters at the molecular level often defies the phase diagrams which dictate the bulk properties. As such, it is conceivable that these new compositions of matter may lead to a wide range of novel materials with unusual properties.

Large *metal alloy clusters* are of potential technological importance in many areas. First, bimetallic clusters are closely related to the important class of *bimetallic* industrial *catalysts*.[61, 62] Second, these new metal alloy clusters may serve as *precursors* to unusual *metal alloy phases, composites*, or *solid-state materials*.[63] Third, metal alloy clusters are ideal models or materials for the studies in *surfaces, interface*, and *interconnection*, all are subjects of great importance in *microelectronics* industry.[64] In fact, metal alloy clusters may find applications in *small-dimension* quantum devices or device fabrications (*nanoelectronics*).[65]

In this review, we shall first discuss a cluster classification scheme which is related to the two broad categories of cluster growth pathways (Section 2). Section 3 describes the structures of a novel series of Au-Ag supraclusters synthesized and structurally characterized in our laboratory. These clusters follow well-defined *design rules*, thereby giving rise to a novel *growth sequence*. The structural systematics of this new class of clusters led to the concept of "*cluster of clusters*" which may be useful in the design, preparation, and characterization of large metal clusters of increasingly high nuclearity via vertex-, edge-, and face-sharing and/or close packing of smaller cluster units as *building blocks*. The observed "cluster of clusters" structures for the Au-Ag supraclusters may also be important in the understanding of the *nucleation* and *growth* of coinage metal particles in solution which is different from that in gas phase (Section 4). In Section 5 we consider the "cluster of clusters" growth in 1-, 2-, and 3-dimensional spaces for *vertex-sharing polyicosahedral* supraclusters. In Section 6, attention is focused on a particular 3-D growth of vertex-sharing polyicosahedral supraclusters which exhibits *self-organization and self-similarity* properties characteristic of the *fractal* behavior often observed in nature. Section 7 summarizes the empirical structural rules. In Section 8, we describe the different metal configurations of the biicosahedral 25-metal-atom supraclusters which led to the concept of "*cluster rotamerism*". The metal core of these clusters may be coined a "*molecular rotor*". Section 9 describes the variegative nature of the satellite bridging halide ligand ring around the "equator" of the biicosahedral rotor. Sections 10 and 11 provide strong structural and electronic evidence for the "cluster of clusters" concept in this series of vertex-sharing polyicosahedral supraclusters. Finally, the novel structure of a 39-atom pure gold cluster with an unusual layer arrangement is described in Section 12. The final Section poses some concluding remarks and suggests a potential usage of the present series of vertex-sharing polyicosahedral supraclusters in general, and the biicosahedral supraclusters in particular, as *molecular machines* and as prototype components in *nanotechnology*.

2 CLUSTER CLASSIFICATION AND CLUSTER GROWTH PATHWAYS

Metal clusters can be categorized into three broad classes as shown in Fig. 1.[66,67] The primary clusters are the simplest *polygonal* or *polyhedral* (or *polytopal* in general) clusters such as a triangular or a tetrahedral cluster, respectively.[68] These clusters, generally of low nuclearity, are of prime importance in that they can be used as *"building blocks"* or *"nucleation core"* for the secondary clusters. The types of secondary clusters can be distinguished on the basis of the *mechanism of cluster growth*: (1) the "cluster of clusters" (COC) pathway[66,67] which gives rise to the "s_n *supraclusters*" via the addition of smaller cluster units as basic "building blocks"; and (2) the "layer-by-layer" (LBL) pathway[66,67] which gives rise to the "v_n *polytopal clusters*" via the addition of successive layers of atoms onto a "nucleation core". An s_n supracluster is defined as a cluster of n smaller cluster units fused together via vertex-, edge-, or face-sharing. Fig. 2 illustrates the early members of supraclusters based on vertex-sharing icosahedral cluster units (each containing 13 atoms). A v_n polyhedral cluster is defined as a cluster with (n+1) atoms on each edge of the polyhedron. Fig. 3 portrays the early members of the v_n icosahedral clusters. The magic numbers, defined as the nuclearity (number of atoms) of a cluster, are given in the parentheses. These magic numbers represent structurally stable, often closed-shell, configurations of atoms in a cluster which are observed experimentally. It should be noted that the LBL growth mechanism encompasses both planar layers and curved shells. The latter may also be coined as the shell-by-shell (SBS) growth pathway. These are related to the "slicing cheese" and "peeling onion" concepts proposed in the literature.[68] The recent discoveries of the onionlike[31] or tubular[32] fullerenes (or a combination thereof) with multi-shell structures provide further support for the LBL or SBS growth pathways.

3 CLUSTER OF CLUSTERS: A NOVEL SERIES OF AU-AG SUPRACLUSTERS

Recently we reported the syntheses and structural systematics of a novel series of Au-Ag supraclusters whose metal frameworks are based on vertex-sharing polyicosahedra.[69-81] In these structures, the basic *building block* is the 13-metal-atom (Au_7Ag_6) icosahedron. The *design rule* is vertex-sharing.[66,67,75-78] We refer to these high-nuclearity metal clusters as *"clusters of clusters"*.[66,67, 69-81] This *"cluster of clusters"* series follows a well-defined *growth sequence* by successive additions of icosahedral units via vertex-sharing.[78] We shall discuss the structures of the early members of this vertex-sharing polyicosahedral supracluster series in the following subsections.

25-Metal-Atom Clusters

The first nontrivial member of this vertex-sharing polyicosahedral cluster series is the 25-metal-atom cluster whose metal core can be considered as two icosahedra sharing a common vertex. Of particular interest is that we have synthesized and structurally characterized a number of 25-metal-atom clusters with distinctly different metal configurations.[69-74] We shall discuss two of them here: $[(p\text{-}Tol_3P)_{10}Au_{13}Ag_{12}Br_8]^+$ **(1)**,[70] and $[(Ph_3P)_{10}Au_{13}Ag_{12}Br_8]^+$ **(2)**.[71] The former has a staggered-staggered-

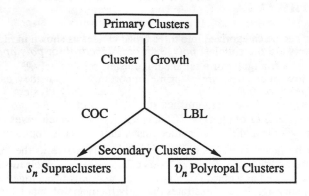

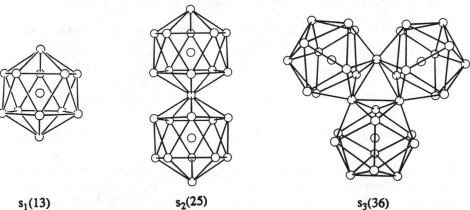

Figure 1   Three classes of clusters: the primary cluster and two types of secondary clusters, s_n supraclusters and υ_n polytopal clusters.

$s_1(13)$ $s_2(25)$ $s_3(36)$

Figure 2   First three members of the s_n supraclusters.

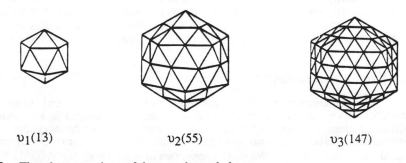

$\upsilon_1(13)$ $\upsilon_2(55)$ $\upsilon_3(147)$

Figure 3   First three members of the υ_n polytopal clusters.

staggered (*sss*) metal configuration between the four metal pentagons whereas the latter can be characterized as staggered-eclipsed-staggered (*ses*), as depicted schematically in Fig. 4. Other 25-metal-atom clusters with intermediate metal configurations will be discussed later. As portrayed in Fig. 5 for **2**, the ten Ph_3P ligands coordinate to ten peripheral (surface) Au atoms in a radial fashion. Of the eight Br ligands, two are terminal, while the remaining six are bridging, connecting the two middle Ag_5 pentagons. The various bridging ligand arrangements will be discussed in a later section.

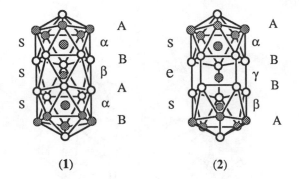

(1) **(2)**

<u>Figure 4</u>   A schematic representation of two distinct metal configurations, staggered-staggered-staggered (*sss*) and staggered-eclipsed-staggered (*ses*), in 25-metal-atom Au-Ag superclusters.

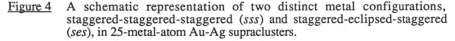

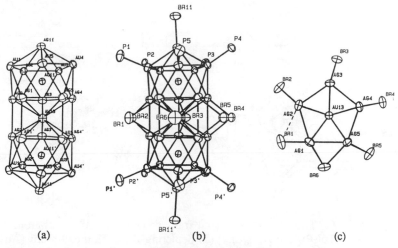

(a) (b) (c)

<u>Figure 5</u>   Molecular architecture of the 25-metal-atom cluster $[(Ph_3P)_{10}Au_{13}Ag_{12} Br_8]^+$ (**2**) as the (SbF_6^-) salt: (a) the *ses* metal core, $Au_{13}Ag_{12}$; (b) the metal-ligand framework, $P_{10}Au_{13}Ag_{12}Br_8$; (c) the projection of the two silver pentagons onto the plane of the six briding bromide ligands as viewed along the idealized 5-fold axis.

38- and 37-Metal-Atom Clusters

The metal framework of the 38-metal-atom cluster $[(p\text{-}Tol_3P)_{12}Au_{18}Ag_{20}Cl_{14}]$ (3)[79] can be described as three 13-atom (Au_7Ag_6) Au-centered icosahedra sharing three Au vertices in a cyclic manner with two capping Ag atoms located on the idealized three-fold axis, as depicted in Fig. 6. The 14 chloride ligands coordinate exclusively to the 20 Ag atoms via three modes of bridging: six doubly bridging, six triply bridging, and two terminal. Globally, the 12 p-Tol$_3$P ligands, which coordinate to the 12 peripheral Au atoms in a radial fashion, form a highly distorted nonbonding twinned cuboctahedron.

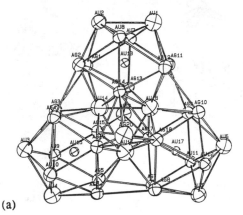

(a)

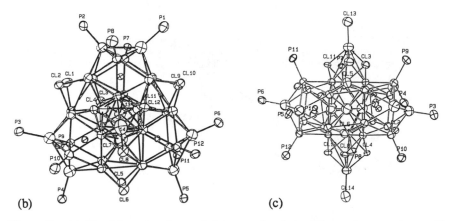

(b)   (c)

Figure 6   (a) The $[Au_{18}Ag_{20}]$ metal framework of the 38-metal-atom cluster $[(p\text{-}Tol_3P)_{12}Au_{18}Ag_{20}Cl_{14}]$ (3) depicting three 13-metal-atom (Au_7Ag_6) Au-centered icosahedra sharing three Au vertices in a cyclic manner plus two capping Ag atoms (Ag19 and Ag20) located on the idealized threefold-fold axis. (b) The $[P_{12}Au_{18}Ag_{20}Cl_{14}]$ framework. (c) Side view of the $[P_{12}Au_{18}Ag_{20}Cl_{14}]$ framework. All radial bonds (12 each) from the centers of the icosahedra (Au13, Au15, and Au17) are omitted for clarity. In (b), Cl13 and Cl14 are designated 13 and 14, respectively, for clarity.

The 37-metal-atom cluster $[(p\text{-}Tol_3P)_{12}Au_{18}Ag_{19}Br_{11}]^{2+}$ **(4)**[80] is related to the 38-metal-atom cluster **(3)** in that one of the two capping Ag moieties, $[AgX_4]^{3-}$, is formally replaced by a triply bridging halide X^- ligand.

46-Metal-Atom Cluster

With 46 metal atoms, $[(Ph_3P)_{12}Au_{22}Ag_{24}Cl_{10}]$ **(5)**[81] represents the largest Au-Ag supracluster. The most obvious description of the structure is as four 13-atom (Au_7Ag_6) Au-centered icosahedra arranged in a tetrahedral arrangement with six shared vertices. As such, the molecular architecture can be described as a *tetrahedron of icosahedra* (see $s_4(46)$ of Fig. 11 in Section 6). The 12 triphenylphosphine ligands are coordinated to the 12 surface Au atoms. The 10 chloride ligands are coordinated to the 24 Ag atoms in the following manner: six doubly bridging and four triply bridging.

4 NUCLEATION AND GROWTH OF COINAGE METAL CLUSTERS

The particular vertex-sharing polyicosahedral growth sequence of 13, 25, 36 (in fact, the closely-related 37 and 38 have been observed), 46, ⋯ metal atoms observed for the Au-Ag clusters may be of importance in the understanding of the nucleation and growth of coinage metal particles in *solution*. In the *gas phase*,[82] on the other hand, the "magic numbers" for the cationic Cu_n^+, Ag_n^+, and Au_n^+ bare clusters occur at n=3, 9, 21, 35, 41, 59, 93, ⋯ whereas that for the anionic Cu_n^-, Ag_n^-, and Au_n^- occur at n=1, 7, 19, 33, 39, 57, 91, ⋯. These observations are consistent with the *jellium* model[83] for the closed-shell electronic configuration of 1s, 1p, 1d, 2s, 1f, 2p, 1g, (2d, 3s, 1h), ⋯, giving rise to total numbers of electrons of 2, 8, 18, 20, 34, 40, 58, 68, 70, 92, ⋯. This difference in the magic number sequence points to the importance of ligation (metal-ligand bonding), and/or crystal effects in dictating the formation and growth of these clusters. Indeed, as will be discussed later, the stereochemistry of the Au-Ag supraclusters is strongly influenced by the ligand environment.

The formation and growth of this series of Au-Ag supraclusters, as elucidated by our systematic structural determination, may represent "snapshots" of the successive stages of the nucleation and growth of ultrafine coinage metal particles (Au and Ag in particular).[78]

5 VERTEX-SHARING POLYICOSAHEDRAL GROWTH

With the 13-atom centered icosahedron as the building block, the supracluster can "grow" into a one-dimensional (1-D) "chain", a two-dimensional (2-D) "sheet", or a three-dimensional (3-D) "solid", as illustrated in Figures 7-9, respectively. Here each icosahedral unit is represented by a sphere. A supracluster with n vertex-sharing icosahedral units is designated as $s_n(N)$ where N is the nuclearity (i.e., the number of metal atoms).

It is important to note that for a polyicosahedral supracluster to "grow", the adjacent icosahedral units must have the *ses* or nearly *ses* metal configuration. As shown in Fig. 4,[73] the relative orientation of the four metal pentagons in $s_2(25)$ can be described as ABAB (*sss*) for **1** and ABBA (*ses*) for **2**, respectively. The corresponding relative orientation of the *two* icosahedra may be described as $\alpha(\beta)\alpha$ and $\alpha(\gamma)\beta$, respectively, where the symbol in the parenthesis represents the orientation of the newly created polyhedron as a result of vertex-sharing. Thus, the former has an "additional" icosahedron in the middle (whose orientation is β, in parenthesis) due to the staggered arrangement of the two middle rings in the *sss* configuration whereas the latter produces a bicapped pentagonal prism whose orientation is designated as γ (in parenthesis).[73]

We note that the *ses* metal configuration allows a "polyicosahedral" growth via vertex-sharing to give bi-, tri-, and tetra-icosahedral supraclusters as exemplified by the structurally characterized 25-(**1** or **2**),[70,71] 38- (**3**)[79] or 37- (**4**),[80] and 46- (**5**)[81] metal-atom Au-Ag supraclusters, respectively. In all these structures, the icosahedral units are linked by (interpenetrating) bicapped pentagonal prisms (γ), instead of the (interpenetrating) bicapped pentagonal antiprisms (viz., icosahedra) (β). The *propagation* of icosahedra via vertex-, edge-, or face-sharing through space is critically dependent upon the relative orientation of the icosahedral units (the building blocks).[73]

In describing the structure and bonding of these supraclusters, $s_n(N)$, it is advantageous to define a *superchain* (1-D), a *superpolygon* (2-D) or a *superpolyhedron* (3-D) formed by the centers of the individual icosahedral units.[75,77,78] The shared vertices are located at the midpoints of the edges of the superchain, the superpolygon, or the superpolyhedron. The structural (atom-counting) rules are given elsewhere in the literature.[77]

For a 1-D growth sequence (Fig. 7), the nuclearity increases as $s_1(13)$ $\rightarrow s_2(25) \rightarrow s_3'(37) \rightarrow s_4''(49)$... by adding one icosahedron at a time since sharing a vertex causes an increment in nuclearity of $(13-1)=12$. The resulting "superchain" may be linear, zig-zag, or a combination thereof (i.e., with kinks or bends). The infinite linear chain analog is observed in the solid-state compounds Ta_2S[84] and Ta_6S.[85] The structural relationship of the high nuclearity clusters and the solid-state compounds therefore provides a link between discrete clusters and the bulk phase.[66]

In the case of 2-D growth, capping an edge of a superpolyhedron increases the nuclearity by $13-2=11$ as exemplified by a series of "two-dimensional" superpolygonal supraclusters: $s_3(36) \rightarrow s_4'(47) \rightarrow s_5''(58) \rightarrow s_6''(69)$ shown schematically in Fig. 8.[66,77] It should be noted, however, that the superpolygons of these supraclusters are puckered instead of planar. The "puckered" nature of these superpolygons produces "corrugated" rather than flat "superpolyhedral sheets".

For 3-D growth, the capping of a triangular face of a superpolyhedron causes the nuclearity to increase by $13-3=10$. As portrayed in Fig. 9,[66,77] the superpolyhedron of the $s_4(46)$ supracluster, for example, is a supertetrahedron. Its capped derivatives (designated by the asterisk to indicate stellation), s_5, s_6^*, s_7^*, and s_8^*, correspond to the mono-, bi-, tri-, and tetra-capped supertetrahedra. Since each cap increases the nuclearity by 10, the corresponding nuclearities are 56, 66, 76, and 86, respectively, which follow the $(10n+6)$ rule.[75]

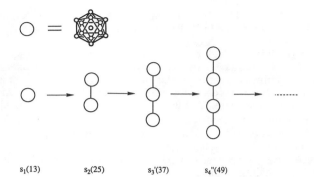

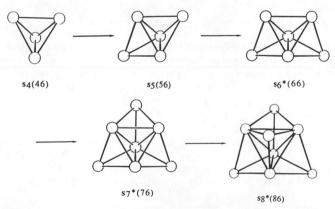

<u>Figure 7</u> One-dimensional growth of the vertex-sharing polyicosahedral supraclusters: $s_1(13)$, $s_2(25)$, $s_3(37)$, s_4 (49), ... ∞, as represented by the growth of a "superchain". Each sphere represents a 13-atom centered icosahedron. The shared vertices are located at the midpoints between the spheres. The nuclearities are in parentheses.

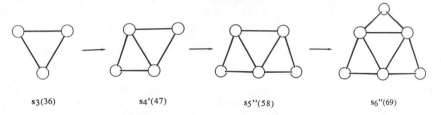

<u>Figure 8</u> Two-dimensional growth of the vertex-sharing polyicosahedral supraclusters: $s_3(36)$, $s_4''(47)$, $s_5''(58)$, and $s_6''(69)$, ..., as represented by mono-, bi-, and tri-edge cappings of the "supertriangle", respectively. Each sphere represents a 13-atom centered icosahedron. The shared vertices are located at the midpoints of the edges. The nuclearities are in parentheses.

<u>Figure 9</u> Three-dimensional growth of the vertex-sharing polyicosahedral supraclusters: $s_4(46)$, $s_5(56)$, $s_6*(66)$, $s_7*(76)$, and $s_8*(86)$, ..., as represented by the successive cappings of the "supertetrahedron" to form mono-, bi-, tri-, and tetracapped supertetrahedra, respectively. Each sphere represents a 13-atom centered icosahedron. The shared vertices are located at the midpoints of the edges. The nuclearities are in parentheses.

6 CLUSTER OF CLUSTERS: A SELF-ORGANIZATION AND SELF-SIMILARITY PRINCIPLE

If the 3-D growth is confined to the surface of a supericosahedron, then the *cluster of clusters* sequence described above for the Au-Ag supraclusters parallels the *atom-by-atom* growth pathway for the primary clusters from a single atom to a 13-atom icosahedron. In other words, if the structures (cf. Fig. 10) of primary clusters $c(n)$ $(n=1\sim13)$ can be considered as models for the "early stages" of cluster growth or particle formation, as envisioned by Briant and Burton,[86] Hoare and Pal,[87] and others, then the structures (cf. Fig. 11) of the Au-Ag supraclusters, $s_n(N)$, of nuclearity (N) ranging from 13 to 127 may be considered as the "intermediate stages" of the cluster growth for Au-Ag supracluster systems where the 13-atom centered icosahedral cluster serves as a basic building block.

In analogy to the primary clusters $c(n)$ (where $n=1\text{-}13$) growth pattern shown in Fig.10, Fig.11 depicts the corresponding supraclusters s_n (N) (where N denotes the nuclearity of the supracluster) formed by n centered icosahedra sharing vertices $(n=1\sim13)$.[78] Here each atom in $c(n)$ is replaced by an icosahedron in $s_n(N)$ with the nuclearity N given by 13n *minus* the number of shared vertices.[75-78] Instead of adding one atom at a time, the supracluster now "grows" by adding one icosahedron at a time, resulting in the formation of the secondary clusters $s_n(N)$.

There is a striking similarity between the early stages of cluster growth with nuclearities ≤ 13 and the intermediate stages with nuclearities ranging from 13 to 127 if one replaces each of the added atoms in the early stages (Fig.10) by a 13-atom icosahedron in the intermediate stages (Fig.11). It is not unreasonable to presume that the former is formed by an atom-by-atom (ABA) growth mechanism whereas the latter is, in fact, a cluster of clusters (COC) growth pathway which capitalizes on the efficiency and geometric design of self-similar modules (icosahedral cluster, $c(13) \equiv s_1(13)$, as the building block).

Since many of the structurally known clusters and supraclusters are formed by *spontaneous self-assembly*, it is concluded that this *similarity* is indeed a manifestation of the *self-organization* and *self-similarity* principle often found in nature.[88,89] The principle of self-organization and self-similarity can lead to large supramolecular assembly. For clusters, *self-organization* means spontaneous assemblage of atoms to form energetically stable clusters of relatively efficient packing and reasonably high symmetry. *Self-similarity* means that the resulting cluster looks more or less alike when examined at different levels of magnification. Indeed, the secondary supraclusters $s_n(N)$ look very much like the corresponding primary clusters $c(n)$ when viewed at roughly half the magnification. Such underlying geometric similarity is often referred to as *fractal*.[88,89]

The spontaneous *self-organization* of *clusters* into *cluster of clusters* (as exemplified by the *self-assembly* of *icosahedra* via vertex-sharing to form, ultimately, an *icosahedron of icosahedra* for the Au-Ag supraclusters) is, in fact, symmetry across scale, pattern within pattern (*viz.* fractal in nature).

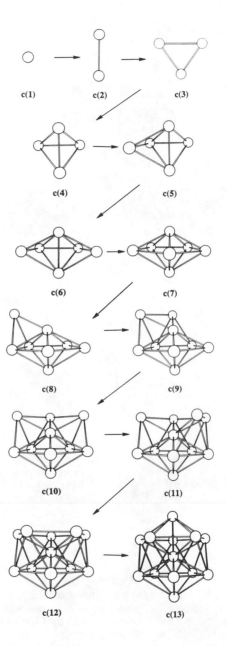

<u>Figure 10</u> Growth sequence of the primary cluster c(n) via an atom-by-atom growth
pathway to form a 13-atom icosahedron where n is the nuclearity.

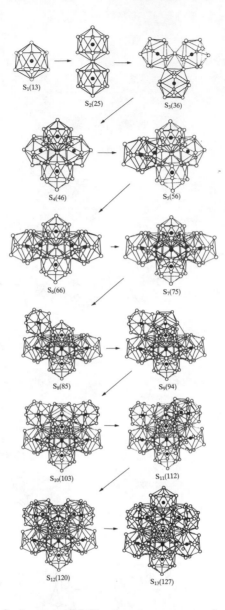

Figure 11 Cluster of clusters (COC) growth sequence from icosahedron to icosahedron of icosahedra. The supraclusters s_n(nuclearity) of n vertex-sharing icosahedra "grow" by adding one icosahedron at a time. The nuclearity (in parenthesis) increases by 13 minus the number of the shared vertices. Each "added" icosahedron is represented by heavy bonds. All radial bonds from the central atom (filled circles) of each icosahedron are omitted for clarity.

7 EMPIRICAL STRUCTURAL RULES

A detailed examination of the stereochemistries of the Au-Ag supraclusters structurally characterized so far revealed six empirical structural rules: (1) the centers of the icosahedra are gold atoms; (2) the "shared" vertices are most likely to be gold atoms; (3) phosphine ligands prefer coordination with surface gold atoms; (4) silver atoms prefer surface sites, especially those at the boundary of neighboring icosahedra; (5) the capping atoms are most likely to be silver atoms; and (6) halide ligands prefer coordination with silver atoms. It is obvious that the first three rules relate to the gold atoms whereas the latter three relate to the silver atoms. These structural rules can be rationalized in terms of the disparity in electronegativity (2.54 for Au vs 1.93 for Ag). The latter is related to relativistic effects.[90,91] Here the more electronegative gold atoms tend to prefer sites of high electron densities, including the centers (Rule 1) of the icosahedra or the shared vertices (Rule 2) whereas the more electropositive silver atoms tend to occupy surface sites (Rule 4) or capping positions (Rule 5). As far as the ligand binding capabilities are concerned, the more electron-donating phosphine ligands prefer more electronegative Au atoms (Rule 3), whereas the more electron-withdrawing halide ligands prefer the more electropositive Ag atoms (Rule 6).

8 CLUSTER ROTAMERISM: DIFFERENT METAL CONFIGURATIONS

As discussed earlier, the adjacent metal pentagons in 25-metal-atom clusters can adopt either the staggered-staggered-staggered (*sss*) configuration, as in **1**, or the staggered-eclipsed-staggered (*ses*) configuration, as in **2**. More recently, we also synthesized and structurally characterized a series of 25-metal-atom clusters with intermediate metal configurations. For example, the metal framework (Fig. 12) of cluster [(p-Tol$_3$P)$_{10}$Au$_{13}$Ag$_{12}$Cl$_8$]$^+$ (**6**)[72] can be described as neither *ses* nor *sss*. In fact, it is nearly halfway between these two extremes. If the four metal pentagons (two outer Au$_5$ rings and two inner Ag$_5$ rings) are projected onto the plane of the bridging ligands, we see a ring of 20 metal atoms more or less evenly distributed around the circle (cf Fig. 13)[72] which is, in reality, a superposition of five pentagons. It is conceivable that in solution, the two metal icosahedra are relatively "free" to rotate (fluxional) about the shared vertex, and that the observed solid-state structure represents a "snapshot" of a "molecular rotor" in motion[72].

Even more interesting is the discovery of two conformers (*ses* and *sss*) for the same monocationic cluster [(Ph$_3$P)$_{10}$Au$_{13}$Ag$_{12}$Br$_8$]$^+$: the *ses* configuration in the SbF$_6^-$ salt (**2**),[71] and the *sss* configuration in the Br$^-$ salt (**7**).[74] The former was discussed earlier (cf. Fig. 5). The latter is portrayed in Fig. 14. This new type of cluster isomerism has been coined "rotamerism".

All of the s$_2$(25) supraclusters discussed so far are monocationic. A dicationic 25-metal-atom cluster, [(p-Tol$_3$P)$_{10}$Au$_{13}$Ag$_{12}$Cl$_7$]$^{2+}$ (**8**),[73] has also been synthesized and structurally characterized by us. As depicted in Fig. 15, the metal configuration is best described as approximately *ses*. It occurs to us that the various metal configurations observed in the solid-state may represent "snapshots" of the many local energy minima of the metal core.

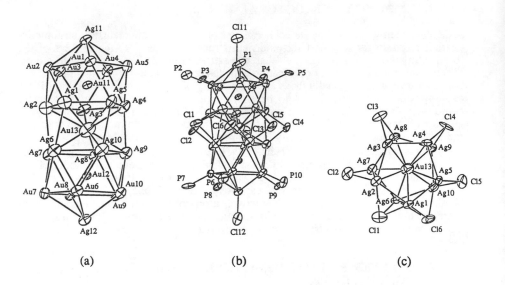

(a)   (b)   (c)

Figure 12   Molecular structure of $[(p\text{-Tol}_3P)_{10}Au_{13}Ag_{12}Cl_8]^+$ (6) (as the PF_6^- salt): (a) biicosahedral framework of the $Au_{13}Ag_{12}$ core; (b) the metal-ligand framework, $P_{10}Au_{13}Ag_{12}Cl_8$; (c) the projection of the two Ag_5 pentagons and the six bridging chloride ligands as viewed along the idealized fivefold axis (which passes through Ag11, Au11, Au13, Au12, and Ag12).

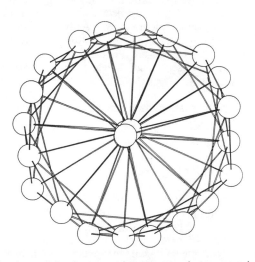

Figure 13   The projection of the four metal pentagons (two outer Au_5 rings and two inner Ag_5 rings) onto the plane of the bridging ligands.

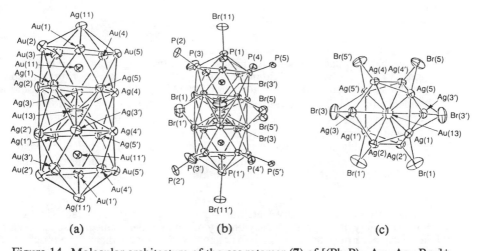

(a)   (b)   (c)

<u>Figure 14</u> Molecular architecture of the *sss* rotamer (**7**) of $[(Ph_3P)_{10}Au_{13}Ag_{12}Br_8]^+$ (as the Br^- salt): (a) the *sss* metal core, $Au_{13}Ag_{12}$; (b) the metal-ligand framework, $P_{10}Au_{13}Ag_{12}Br_8$; (c) the projection of the two silver pentagons onto the plane of the six bridging Br ligands. The symmetry-related atoms are designated as primes. The metal core has an idealized fivefold rotation symmetry (passing through Ag11, Au11, Au13, Au1', and Ag11'). All radial bonds (12 each) from Au11 and Au11' have been omitted for clarity. The *ses* rotamer (**2**) of the same cluster is portrayed in Figure 5.

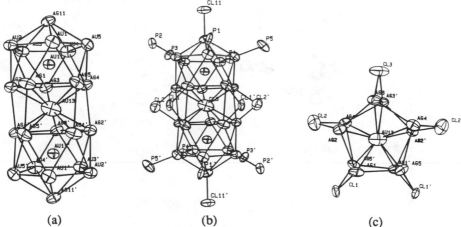

(a)   (b)   (c)

<u>Figure 15</u> Molecular architecture of $[(p\text{-}Tol_3P)_{10}Au_{13}Ag_{12}Cl_7]^{2+}$ (**8**) (as the SbF_6^- salt): (a) the metal core, $Au_{13}Ag_{12}$; (b) the metal-ligand framework, $P_{10}Au_{13}Ag_{12}Cl_7$; (c) the projection of the two silver pentagons onto the plane of the five doubly bridging chloride ligands as viewed along the idealized fivefold axis which passes through Ag11, Au11, Au13, Au11', and Ag11'. Atoms related by the crystallographic twofold (C_2-2) symmetry are designated as primes. All radial bonds (12 each) from Au11 and Au11' have been omitted for clarity.

9 VARIEGATIVE BRIDGING LIGAND ENVIRONMENTS

A detailed analysis of the structures of these 25-metal-atom Au-Ag clusters (**2, 6, 7,** and **1**) revealed a stellated ring of six halide ligands bridging the two inner silver pentagons. Conceptually, the six bridging halide ligands form a "satellite ring" around the "equator" of the biicosahedral rotor. For example, the bridging halide ligand arrangements in **2, 6, 7** and **1** can be described as *ortho*-qq'd$_4$,[71] *ortho*-t$_2$d$_4$,[72] *para*-t$_2$d$_4$,[74] and *para*-d$_2$t$_4$,[70] respectively, as illustrated schematically in Fig. 16.[74] Here d, t, and q designate doubly, triply, and quadruply bridging, respectively. (The prime indicates highly asymmetrical bridging). Instead of a stellated ring of six halide ligands bridging the two inner silver pentagons, the dicationic supracluster **8**[73] has only five doubly-bridging chloride ligands linking the two Ag$_5$ pentagons (cf. Fig. 15(c)).

Apparently such a variegative satellite ring of bridging ligands is caused by the incommensurability of the pentagonal symmetry of the two five-membered silver rings and the hexagonal arrangement of the six halogen ligands bridging them. The observed bridging arrangements, which often give rise to highly asymmetric bridging modes, represent a compromise between the attractive silver-halide interactions and the repulsive halide···halide (van der Waals) interactions.

  ortho - qq'd$_4$ *ortho* - t$_2$d$_4$ *para* - t$_2$d$_4$ *para* - d$_2$t$_4$

  2   **6**   **7**   **1**

Figure 16   The projection of the two silver (small circles) pentagons onto the plane of the six bridging halide (large circles) ligands of **2, 6, 7** and **1** (see text). The symbols d, t, and q designate doubly, triply, and quadruply bridging, respectively. The prime indicates highly asymmetrical bridging.

10 STRUCTURAL EVIDENCE OF "CLUSTER OF CLUSTERS" CONCEPT

The observation of various metal configurations between *ses* and *sss* for the metal core of the 25-metal-atom clusters provides strong evidence for the concept of *cluster of clusters*, as applied to this series of vertex-sharing polyicosahedral Au-Ag supraclusters. Further structural evidence comes from the interpentagonal separations with the intraicosahedra (outer) Au_5-Ag_5 interplanar distances being 0.5Å shorter than the intericosahedra (inner) Ag_5-Ag_5 interplanar distance.[69-74] Moreover, the average intra-icosahedral metal-metal distances are significantly (ca 0.1Å) shorter than the average intericosahedral distances.[69-74] These observations suggest that intraicosahedral bonding is substantially stronger than intericosahedral bonding, thereby reinforcing the *cluster of clusters* concept[66,67,69-81] in which the individual icosahedron serves as the basic building block.

11 ELECTRONIC EVIDENCE OF "CLUSTER OF CLUSTERS" CONCEPT

The electronic requirements of the vertex-sharing polyicosahedral Au-Ag supraclusters have been theoretically defined by us on the basis of the "cluster of clusters" (C^2) model.[75,76] In fact, the agreement between the experimentally observed electron counts and those predicted by the C^2 model provides strong electronic evidence for the "cluster of clusters" concept for this series of supraclusters.[75]

For the 25-atom cluster formed by two icosahedra sharing one vertex, the C^2 model predicts B=2x13 (icosahedron)-(1x3)(sharing one vertex)=23 skeletal electron pairs, $T=6V_m+B=6x(25-2)+23=161$ total electron pairs or a total valence electron count of N=2x161=322. One example is the $[(p\text{-}Tol_3P)_{10}Au_{13}Ag_{12}Br_2(\mu\text{-}Br)_2(\mu_3\text{-}Br)_4]^+$ cluster (**1**)[70] for which the N_{obs} of (10x2+25x11+2x1+2x3+4x5-1)=322 valence electrons is in accordance with the calculated value. Likewise, three 13-atom icosahedra sharing three vertices in a cyclic manner results in B=(3x13)-(3x3)=30 pairs for $s_3(36)$. For such a 36-metal atom cluster, $T=6V_m+B=6x(36-3)+30=228$ electron pairs and N=2x228=456 electrons. This cluster is presently unknown. However, the closely related 38-metal atom cluster $[(R_3P)_{12}Au_{18}Ag_{20}Cl_{14}]$ (**3**)[79] (where R=p-MeC_6H_4) is predicted to have 456+2x18=492 valence electrons if the two exopolyhedral Ag atoms do not form metal-metal bonds with the 36-atom triicosahedral framework, as is indeed observed (N_{obs}=12x2+38x11+2x1+6x3+6x5=492 electrons). It is predicted that if six metal-metal bonds (three on each side) are formed between the two exopolyhedral Ag atoms and the polyhedral atom framework, the electron count would be 492-12=480 electrons. As described elsewhere in the literature,[75] the number of skeletal electron pairs for an $s_n(N)$ supracluster with n vertex-sharing icosahedral units can also be given by[92]

$$B_n = 4n + 18 \qquad (1)$$

and the total number of electron pairs by

$$T_n = 58n + 54 \tag{2}$$

In terms of the total number of valence electrons,

$$N_n = 116n + 108 \tag{3}$$

We can use eq. 1-3 to rationalize or predict the electron counts of vertex-sharing polyicosahedral supraclusters.[75]

It should also be pointed out that the 4n term in eq. 1 may be interpreted as the "polyoctet rule" in that each icosahedron in a polyicosahedral supracluster contributes eight electrons (i.e., four electron pairs) to the B value:[75]

$$B_p = 4n \tag{4}$$

(Here we ignore the ten d electrons per metal atom). If we assume that each coinage metal atom contributes one s electron (as a pseudo hydrogen) to, and that each halide ligand withdraws one electron from, the cluster, then B_p can be calculated by

$$B_p = (M - X - Q) / 2 \tag{5}$$

where M, X, and Q refer to the numbers of metal atoms, halide ligands, and the overall charge, respectively. For example, the 25-metal-atom cluster $[(p\text{-}Tol_3P)_{10}Au_{13}Ag_{12}Br_8]^+$ (**1**)[70] has $B_p=(25\text{-}8\text{-}1)/2=8$ electron pairs as predicted for a biicosahedral supracluster (4n=4x2=8 electron pairs). Similarly, for the 25-metal-atom dicationic cluster, $[(p\text{-}Tol_3P)_{10}Au_{13}Ag_{12}Cl_7]^{2+}$ (**8**),[73] $B_p=(25\text{-}7\text{-}2)/2=8$ electron pairs, also as expected.

For the 38-metal-atom cluster, $[(p\text{-}Tol_3P)_{12}Au_{18}Ag_{20}Cl_{14}]$ (**3**),[79] with three icosahedra sharing three vertices, $B_p=(38\text{-}14)/2=12$ electron pairs, which agrees with the triicosahedral model (4n=4x3=12 electron pairs). Similarly, the 37-metal-atom cluster, $[(p\text{-}Tol_3P)_{12}Au_{18}Ag_{19}Br_{11}]^{2+}$ (**4**),[80] has $B_p=(37\text{-}11\text{-}2)/2=12$ pairs of electrons, once again, as expected.

12 PURE GOLD CLUSTER

With only a few exceptions (e.g., Mn), nearly all pure metals crystallize in one of the three basic close-packing structures: face-centered cubic (fcc), hexagonal closed-packing (hcp), and body-centered cubic (bcc). In the "cluster phase", constraints of the infinite lattice are lifted such that the metal arrangements can adopt any one of the close-packing structures,[93-97] or some combinations and/or variants thereof (such as pentagonal or icosahedral packing), depending upon the electronic and stereochemical requirements of the metal core and the ligand environment.[66,67] The structures of these clusters can often be described as LBL or SBS. For example, $[Rh_{13}(CO)_{24}H_{5-q}]^{q-}$ [93] has a 3:7:3 layered hcp structure whereas $[Pt_{38}(CO)_{44}]^{2-}$ [94] has a 7:12:12:7 layered fcc structure. $[Rh_{15}(CO)_{27}]^{3-}$ [95] and $[Rh_{22}(CO)_{37}]^{4-}$, [96] on the other hand, have mixed bcc/hcp and fcc/hcp structures, respectively. We recently reported a novel pure gold cluster $[(Ph_3P)_{14}Au_{39}Cl_6]^{2+}$ (9)[98] which has an unprecedented 1:9:9:1:9:9:1 layered hcp/hcp' structure (Fig. 17). In one-half of the molecule, the layered structure 1:9:9:1 can be described as nearly hexagonal close-packing (hcp) ABA layering (except that Au10 is displaced toward the center of the cluster). The two halves of the cluster (layers ABA and A'B'A') are twisted by 30°, creating a hexagonal antiprismatic hole in which the interstitial gold atom (Au10) resides. This "interface" may be characterized as a "twist dislocation" which is describable as a ABA(B)ABA hcp arrangement with the central layer(B) missing, followed by a 30° twist of the two halves of the molecule about the idealized threefold axis.

Of the 39 Au atoms, only *one* (the central Au atom) can be considered as a "bulk" atom (completely encapsulated, interstitial atom). The 38 "surface" Au atoms can be categorized into three types: 14 of them are coordinated by phosphine ligands, 6 by chloride ligands, and the remaining 18 are somewhat *recessed* and *uncoordinated* by ligands. We note that unligated metal atoms on the surface of a cluster are rather unusual and may be important in the understanding of surfaces, interfaces, colloids, catalysts, etc.[98]

The central atom, Au10, which resides at the center of the hexagonal antiprismatic hole, has 12 Au-Au contacts (average distance 3.040(8)Å) with the two gold hexagons. If we include six additional Au-Au distances at an average value of 3.757(8)Å from layer B, the central cage can also be described as two half cuboctahedra fused with a hexagonal antiprism by sharing hexagonal faces (10), as depicted in Fig. 17(d). In fact, this 18-vertex polyhedron is one of the 92 possible convex polyhedra with regular faces.

13 CONCLUDING REMARKS AND FUTURE PROSPECTS: MOLECULAR MACHINES AND NONATECHNOLOGY

Recent developments in cluster synthesis have produced many high nuclearity metal clusters of sizes approaching that of small particles. Some of these clusters have metal arrangements resembling fragments of metallic lattices and thus may be considered as a miniature bulk. Yet, others have no structural features in common with that of the bulk. These metal clusters of definitive size and shape provide an opportunity for the study of the evolution of band structure from atomic to molecular to bulk.

Our work in this area gives rise to a series of phosphine Au-Ag halide supraclusters whose structures are based on *vertex-sharing polyicosahedra*. These clusters follow well-defined *design rules* and a novel *growth sequence*. The structural systematics of this new class of clusters led to the concept of "*cluster of clusters*" which may be useful in the design, preparation, and characterization of large metal clusters of increasingly high nuclearity via vertex-, edge-, and face-sharing and/or close packing of smaller cluster units as *building blocks*. The observed polyicosahedral structures for the Au-Ag supraclusterrs may also be important in the understanding of the *nucleation* and *growth* of coinage metal particles in solution which is different from that in gas phase.   In this review, we consider the "cluster of clusters" growth in 1-, 2-, and 3-dimensional spaces for *vertex - sharing poly-*

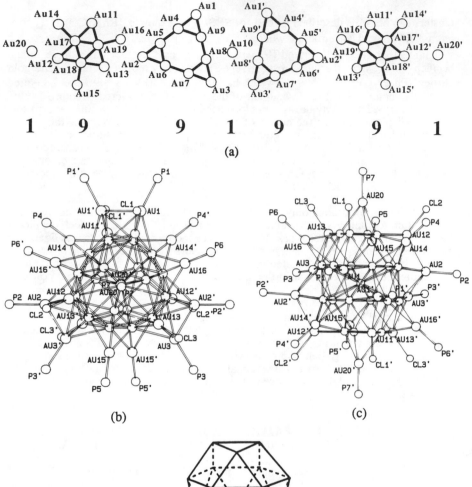

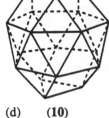

Figure 17 Molecular architecture of the 39-atom pure gold cluster $[(Ph_3P)_{14}Au_{39}Cl_6]^{2+}$ (9): (a) individual 1:9:9:1:9:9:1 metal layers viewed along the idealized threefold axis; (b) the metal-ligand framework, $P_{14}Au_{39}Cl_6$, as viewed along the idealized threefold axis; (c) side view of $P_{14}Au_{39}Cl_6$ as projected along the crystallographic twofold symmetry axis; (d) a schematic representation of the 18-vertex polyhedral cage (10) in which the unique interstitial atom (Au10) resides. The twofold symmetry-related atoms are designated as primes.

icosahedral supraclusters. Attention is focused on a particular 3-D growth sequence which exhibits *self-organization and self-similarity* properties characteristic of the *fractal* behavior often observed in nature. Finally, the novel structure of a 39-atom pure gold cluster with an unusual layer arrangement is described.

The local fivefold symmetry in the vertex-sharing polyicosahedral sequence of our Au-Ag supraclusters system is of relevance to other research areas of current interest. For example, fivefold symmetry, which is not allowed crystallographically, has recently been found in quasicrystals.[99-101] The latter have long-range orientational order but only quasiperiodic translational order. Pentagonal or icosahedral packing (*pip*) are also quite common in clusters and fine particles[34-37] where the crystallographical constraints are partially or completely lifted. *Pip* has also been implicated in the early stages of cluster growth[86,87] and as structural units in amorphous materials.[102]

Miniaturization poses a major challenge in the development of nanoelectronics.[65] Further progress must rely on the fabrication of components in the nanometer range. "Nanotechnology" is the culmination of many facets of developments in the nanorealm, including nanofabrication, nanomechanics, nanostructures, quantum devices, molecular machines, molecular computers, etc.[103,104] Two approaches for building such devices can be distinguished. The "top-down" method is a "scaling down" of the present state of the art microelectronics technology. The "bottom-up" approach seeks to build molecular or supramolecular assemblies with specific functions and properties from smaller building blocks.[105-108] These organized and functional supramolecules[105] can be constructed either by direct synthesis or by self-assembly. These supramolecules are prototypes for the construction of more sophisticated molecular devices.[106] The "cluster of clusters" concept developed by us[66,67,69-81] for building large cluster assemblies is useful in the design and synthesis of "supraclusters" which may ultimately serve as models for the building of *molecular machines* of well-defined structures. The driving force for such spontaneous cluster assemblage is the "self-organization self-similarity" principle (fractal) often observed in nature.[78,88,89] In this regard, the new series of 25-metal-atom-cluster described in this review can be visualized as a "molecular rotor"[72,74] with two metal icosahedral units sharing a vertex. The ellipsoidal (prolate) metal core measures approximately $1 \times 1 \times 1.5 nm^3$. Taken together, this series of supraclusters can be envisaged as a prototype for the fabrication of molecular mechanical rotary devices which may be of potential use in *nanotechnology.*

Acknowledgment -Acknowledgment is made to the National Science Foundation (CHE-9115278) for financial support of this research. BKT would also like to acknowledge the important contributions of the present and past members of his group whose names appear in the references.

REFERENCES
1. Cotton,F. A., and Wilkinson, G. *"Advanced Inorganic Chemistry "*(Fifrth Edition), **1989**.
2. Echt, O., Sattler, K., Recknagel, E. *Phys. Rev. Lett.* **1981**, *47*, 1121.
3. Ozin, G. A.; Mitchell, S.A. *Angew. Chem. Int. Ed. Engl.* **1983**, *22*, 674.
4. Dietz, T.G.; Duncan, M.A.; Power, D.E.; Smalley, R.E. *J. Chem. Phys.* **1981**, *74*, 6511.
5. Rohlfing, E.A.; Cox, D. M.; Kaldor, A. *Chem. Phys. Lett.* **1983**, *99*, 161.
6. Kroto, H.W.; Heath, J.R.; O'Brien, S.C.; Curl, R.F.; Smalley, R.E. *Nature* (London) **1985**, *318*, 162.

7.   Martin, T. P. *Festkorperprobleme* XXIV, **1984**, 1.
8.   de Heer, W.A.; Knight, W.D. *"Elemental and Molecular Clusters"* Spinger Series 6, Spinger-Verlag, **1988**.
9.   Smalley, R.E. *Springer Ser. Opt. Sci.* **1985**, *49*, 317.
10. Kharas, K. and Dahl, L. *Adv. Chem. Phys.* **1988**, *70(2)*, 1.
11. (a) Chini, P. *Gazz. Chim. Ital.* **1979**, *109*, 225; (b) Chini, P., Longoni, G., Albano, V. G. *Adv. Organomet. Chem.* **1976**, *14*, 285; (c) Chini, P. *J. Organomet. Chem.* **1980**, *200*, 37.
12. Shriver, D. F.; Kaesz, H. D.; Adams, R. D. Ed., *"The Chemistry of Metal Cluster Complexes"*, VCH Publishers, New York, **1990**.
13. (a) Johnson, B. F. G., Ed., *"Transition Metal Clusters"*, Wiley-Inter-science, Chichester, England, **1980**; (b) Johnson, B.F. G.; Benfield, R.E. in *"Topics in Inorganic and Organometallic Stereochemistry"* (Edited by Geoffroy G.), **1981**.
14. Moskovits, M., Ed., *"Metal Clusters"*, Wiley-Interscience, New York, **1986**.
15. Fenske, D.; Ohmer, J.; Hachgenei, J.; Merzweiler, K. *Angew. Chem. Int. Ed. Engl.* **1988**, 27, 1277.
16. Schmid, G. *Struct. Bonding* (Berlin) **1985**, *62*, 51.
17. Vahrenkamp, H. *Struct. Bonding* (Berlin), **1977**, *32*, 1.
18. Chisholm, M. H. *ACS Symp. Ser.* **1981**, *155*, 17.
19. Ozkar, S., Ozin, G. A., Moller, K., Bein, T. *J. Am. Chem. Soc.* **1990**, *112*, 9575.
20. Herron, N., Wang, Y., Eddy, M., Cox, D., Moller, K., Bein, T. *J. Am. Chem. Soc.* **1989**, *111*, 530.
21. Stucky, G. D., MacDougall, J. E. *Science* **1990**, *247*, 669.
22. Kim, Y.; Seff, K. *J. Am. Chem. Soc.* **1978**, *100*, 175.
23. Corbett, J.D. *Prog. Inorg. Chem.* **1976**, *21*, 129.
24. Schnering, H.G.von *Angew. Chem. Int. Ed. Engl.* **1981**, *20*, 33.
25. Simon, A. *Angew. Chem. Int. Ed. Engl.* **1981**, *20*, 1.
26. Krebs, B.; Hucke, M.; Brendel, C. S. *Angew. Chem., Int. Ed. Engl.* **1982**, *21*, 445.
27. Belin, C. H. E.; Corbett, J. D.; Cisar, A *J. Am. Chem. Soc.* **1977**, *99*, 7163.
28. Briant, C. E.; Theobald, B. R. C.; White, J. W.; Bell, L. K.; Mingos, D. M. P.; Welch, A. J. *J. Chem. Soc., Chem. Comm.* **1981**, 201.
29. Washecheck, D. M.; Wucherer, E. J.; Dahl, L. F.; Ceriotti, A.; Longoni, G.; Manassero, M.; Sansoni, M.; Chini, P. *J. Am. Chem. Soc.* **1979**, *101*, 6110.
30. (a) Kroto, H. W.; Heath, J. R.; O'Brien, S. C.; Curl, R.F.; Smalley, R. E. *Nature*, **1985**, 162; (b) Kratschmer, W.; Lamb, L.D.; Fostiropoulus, K.; Huffman, D.R. *Nature*, **1990**, 354.
31. Ugarte, D. *Nature*, **1992**, *359*, 707.
32. Iijima, S. *Nature*, **1991**, *354*, 56.
33. Castleman, A. W., Jr.; Keesee, R.G. *Acc. Chem. Res.* **1986**, *19*, 413.
34. Renou, A.; Gillet, M. *Surf. Sci.* **1981**, *106*, 27 and references cited therein.
35. (a) Burton, J. J. *Cat. Rev.* **1974**, *9*, 209; (b) Hoare, M. R.; Pal, P. *Adv. Phys.* **1971**, *20*, 161.
36. Klabunde, Kenneth, J. *"Chemistry of Free Atoms and Particles"* Academic Press, New York, NY **1980**.
37. Baetzold, R. C. *J. Chem. Phys.* **1971**, *55*, 4363.
38. (a) Boyle, P.D.; Johnson, B.J.; Buehler, A.; Pignolet, L. H. *Inorg. Chem.* **1986**, *25*, 5; (b) Alexander, B. D.; Boyle, P. D.; Johnson, B. J.; Casalnuovo, A. L.; John, S. M.; Mueting, A. M.; Pignolet, L. H. *Inorg. Chem.* **1987**, *26*, 2547.
39. (a) Ito, L. N.; Sweet, D. J.; Mueting, A. M.; Pignolet, L. H.; Steggerda, J. J.; Schoondergang, M. F. J. *Inorg. Chem.* **1989**, *28*, 3696; (b) Kanters, R. P. F.; Bour, J. J.; Schlebos, P. P. J.; Bosman, W. P.; Behm, H.; Steggerda, J. J.; Pignolet, L. H. *Inorg. Chem.* **1989**, *28*, 2591; (c) Bour, J. J.; Kanters, R. P. F.; Schlebos, P. P. J.; Steggerda, J. J. *Recl. Trav. Chim. Pays-Bas* **1988**, *107*, 211; (d) Kanters, R. P. F.; Schlebos, P. P. J.; Bour, J. J.; Bosman, W. P.; Behm, H.; Steggerda, J. J.; *Inorg. Chem.* **1988**, *27*, 4034; (e) Kanters, R. P. F.; Schlebos, P. P. J.; Bour, J. J.; Bosman, W. P.; Smits, J. M. M.; Beurskens, P. T.; Steggerda, J. J.; *Inorg. Chem.* **1990**, *29*, 324.
40. Hayward, C. M. T., Shapley, J. R., Churchill, M. R., Bueno, C. and Rheingold, A. L. *J. Am. Chem. Soc.* **1982**, *104*, 7347.
41. (a) Amoroso, A. J.; Gade, L. H.; Johnson, B. F. G.; Lewis, J. Raithby, P. R.; Wong, W.-T. *Angew. Chem. Int. Ed. Engl.* **1991**, *30*, 107; (b) Jackson, P. F., Johnson, B. F. G., Lewis, J., Nelson, W. J. H., McPartlin, M. *J. Chem. Soc., Dalton Trans.* **1982**, 2099; (c) Drake, S. R.; Henrick, K.; Johnson, B. F. G.; Lewis, J.; McPartlin M.; Morris, J. *J. Chem. Soc., Chem. Comm.* **1986**, 928.
42. (a) Broach, R. W., Dahl, L. F., Longoni, G., Chini, P., Schultz, A. J., Williams, J. M. *Adv. Chem. Ser.* **1978**, *No. 167*, 93; (b) Longoni, G.; Dahl, L.; et al unpublished results.
43. (a) Ceriotti, A, Demartin, F, Longoni, G., Manassero, M., Marchionna, M., Piva, G., Sansoni, M. *Angew. Chem. Int. Ed. Engl.* **1985**, *24*, 697; (b) Ceriotti, A.; Fait, A.; Longoni, G.; Piro, G. *J. Am. Chem. Soc.* **1986**, *108*, 8091; (c) Longoni, G.; Manassero, M.; Sansoni, M. *J. Am. Chem. Soc.* **1980**, *102*, 3242; (d) Fumagalli, A., Martinengo, S., Ciani, G. and Sironi, A. *J. Chem. Soc., Chem. Commun.* **1983**, 453.
44. (a) Mednikov, E. G.; Eremenko, N. K.; Slovokhotov, Yu. L.; Struchkov, Yu. T. *J. Chem. Soc., Chem. Comm.* **1987**, 218; (b) Mednikov, E. G.; Eremenko, N. K. *J. Organomet. Chem.* **1986**, *C35-C37*, 301.
45. (a) Fenske, D.; Krautscheid, H. *Angew. Chem. Int. Ed. Engl.* **1990**, *29*, 1452; (b) Fenske, D.; Ohmer, J.; Hachgenei, J. *Angew. Chem. Int. Ed. Engl.* **1985**, *24(11)*, 993.
46. Payne, M. W.; Leussing, D. L.; Shore, S. G. *J. Am. Chem. Soc.* **1987**, *109*, 617.
47. (a) Braunstein, P.; Rose, J.; *Stereochem. Organomet. Inorg. Compounds* **1988**, *3*, 320; (b) Gladfelter, W. L.; Geoffroy, F. L. *Adv. Organomet. Chem.* **1980**, *18*, 207.

48. (a) Callahan, K. P.; Hawthoron, M. F. *Adv. Organomet. Chem.* **1976**, *14*, 145; (b) Grimes, R. N. *Pure Appl. Chem.* **1987**, *59*, 847; (c) Amini, M.M.; Fehlner, T. P.; Long, G. J.; Politowski, M. *Chem. of Materials* **1990**, 2, 432.
49. Bradley, J. S. in Moskovits, M. ed., *"Metal Clusters"* Wiley, New York, **1986**, Ch.5, p.105.
50. (a) Willett, R. D. *J. Coord. Chem.* **1988**, *19*, 253; (b) Koenig, T. W.; Hay, B. P.; Finke, R. G. *Polyhedron* **1988**, 7, 1479.
51. (a) Whitmire, K. H. *J. Coord. Chem.* **1988**, *17*, 95; (b) Ichikawa, M. *Chem.Tech.* **1982**, 674.
52. (a) Lauher, J. W., *J. Am. Chem. Soc.*, **1978**, *100*, 5305; (b) Lauher, J. W., *ibid*, **1979**, *101*, 2604; (c) Lauher, J. W., *J. Organomet. Chem.*, **1981**, *213*, 25.
53. (a) King, R. B., *Inorg. Chim. Acta*, **1986**, *116*, 99; (b) King, R. B., *ibid.*, **1986**, *116*,109; (c) King, R. B., *ibid.*, **1986**, *116*, 119; (d) King, R. B., *ibid.*, **1986**, *116*, 125; (e) King, R. B.; Rouvray, D. H., *J. Am. Chem. Soc.*, **1977**, *99*, 7834.
54. See, for example, Wertheim, G.K.; Kwo, J.; Teo, B.K.; Keating, K.A. *Solid State Comm.* **1985**, *55(4)*, 357.
55. Monot, R.; Narbel, C.; Borel, J.-P *Cimento Ital. Fis., B* **1974**, *19*, 253.
56. Borel, J.-P; Millet, J.-L *J. Phys. Colloq.* **1977**, *C2*, 115.
57. Knight, W. D. *J. Phys. Colloq.* **1977**, *C2*, 109.
58. Rossetti, R.; Nakahara, S.; Brus, L. E. *J. Chem. Phys.* **1983**, *79*, 1086.
59. (a) Kubo, R. *J. Phys. Soc. Japan* **1962**, *17*, 975; (b) Kubo, R.; Kawabata, A.; Kobayashi, S. *Ann. Rev. Mater. Sci.* **1984**, *14*, 49.
60. Messmer, R. P.; Knudson, S. K.; Johnson, K. H.; Diamond, J. B.; Yang, C. Y. *Phys. Rev. B* **1976**, *13*, 1396.
61. (a) Sinfelt, J. H. "Bimetallic Catalysts: Discoveries, Concepts, and Applications", J. Wiley & Sons, New York, **1983**, pp. 1-164, and references cited therein. (b) Sachtler, W. M. H.; Van Santen, R. A. *Adv. Catal.* **1977**, *26*, 69.
62. Gates, B. C.; Guczi, L.; Knozinger, H. *"Metal Clusters in Catalysis"*, Elsevier Ansterdem, **1988**.
63. For leading references, see, for example, Burdett, J. K. *Prog. Solid State Chem.* **1984**, *15*, 173.
64. For an interesting historical review of microelectronics, see T. R. Reid, *"The Chip"*, Simon and Schuster, New York (**1984**), p. 195.
65. (a) Moffat, A. S. *Mosaic*, **1990**, *21*, 30; (b) "Physics and Technology of Submicron Structures", Heinrich, H.; Bauer, G.; Kuchar, F.; Eds., Springer-Verlag, Berlin, Heidelberg, **1988**.
66. Teo, B. K., Zhang, H. *J. Cluster Science*, **1990**, *1*, 155.
67. (a) Teo, B. K., Zhang, H. *Polyhedron,* **1990**, *9*, 1985; (b) Teo, B. K. *Polyhedron* **1988**, *7*, 2317.
68. (a) Teo, B. K.; Sloane, N. J. A., *Inorg. Chem.*, **1985**, *24*, 4545; (b) Sloane, N. J. A.; Teo, B. K., *J. Chem. Phys.*, **1985**, 83, 6520; (c) Teo, B. K.; Sloane, N. J. A., *Inorg._Chem.*, **1986**, *25*, 2315.
69. Teo, B. K.; Keating, K. *J. Am. Chem. Soc.* **1984**, *106*, 2224.
70. Teo, B. K.; Zhang, H.; Shi, X. *Inorg. Chem.* **1990**, *29*, 2083.
71. Teo, B.K.; Shi, X.; Zhang, H. *J. Am. Chem. Soc.*, **1991**, *113*, 4329.
72. Teo, B. K.; Zhang, H *Angew. Chem. Int. Ed. Engl.* **1992**, *31*, 445.
73. Teo, B. K.; Zhang, H *Inorg. Chem.* **1991**, *30*, 3115.
74. Teo, B.K.; Shi, X.; Zhang, H. *J. Chem. Soc., Chem. Commun.* **1992**, 1195.
75. Teo, B.K.; Zhang, H. *Inorg.Chem.* **1988**, *27*, 4141.
76. Teo, B.K.; Zhang, H. *Inorg. Chim. Acta* **1988**, *144*, 173.
77. Teo, B.K., Zhang, H. *J. Cluster Science*, **1990**, *1*, 223.
78. Teo, B.K.; Zhang, H. *Proc. Nat. Acad. Sci.* **1991**, *88*, 5067.
79. (a) Teo, B. K.; Zhang, H.; Shi, X. *J. Am. Chem. Soc.*, **1990**, *112*, 8552; (b) Teo, B. K.; Hong, M., Zhang, H., Huang, D.; Shi, X. *J. Chem. Soc., Chem. Commun.* **1988**, 204.
80. Teo, B. K., Hong, M. C., Zhang, H., Huang, D. B. *Angew. Chem., Int. Ed. Engl.* **1987**, *26*, 897.
81. Teo, B.K.; Shi, X.; Zhang, H. quoted in *Chem. Eng. News*, **1989**, *67*, 6.
82. Katakuse, I.; Ichihara, T.; Fujita, Y.; Matsuo, T.; Sakurai, T.; Matsuda, H. *Inter. J. Mass. Spect. and Ion Proc.* **1986**, *74*, 33.
83. Knight, W. D.; Clemenger, K.; de Heer, W. A.; Saunders, W. A.; Chou, M. Y.; Cohen, M. L. *Phys. Rev. Lett.* **1984**, *52*, 2141.
84. Franzen, H. F.; Smeggil, J.G. *Acta Crystallogr.* **1969**, *B25*, 1736.
85. Franzen, H.F.; Smeggil, J.G. *Acta Crystallogr.* **1970**, *B26*, 125.
86. Briant, C. L.; Burton, J. J. *Phys. Status Solidi*, **1978**, *85*, 393.
87. Hoare, M. R.; Pal, P. *Adv. Phys.*, **1971**, *20*, 161.
88. Mandelbrot, B.B. (**1983**) *The Fractal Geometry of Nature* (W.H. Freeman & Co., New York).
89. Peitgen, H.O.; Richter, P. (**1986**) *The Beauty of Fractals* (Springer-Verlag, Heidelberg).
90. Pyykko, P.; Desclaux, J. *Acc. Chem. Res.*, **1979**, *12*, 276.
91. Pitzer, K.S. *Acc. Chem. Res.*, **1979**, *12*, 271.
92. For n=2, eq. 1-3 become $B_n = 4n+15$, $T_n = 58n+45$ and $N_n = 116+90$.
93. Ciani, G.; Sironi, A.; Martinengo, S. *J. Chem. Soc. Dalton.* **1981**, 519.
94. Longoni, G. and Dahl, L., et al unpublished results quoted in Kharas, K. and Dahl, L. F. *Adv. Chem Phys.* **1988**, 70,1.
95. Martinengo, S.; Ciani, G.; Sironi, A.; Chini, P. *J. Am. Chem. Soc.* **1978**, *100*, 7096.
96. Vidal, J.L.; Schoening, R.C.; Troup, J.M. *Inorg. Chem.*, **1981**, *20*, 227.
97. Teo, B. K. *J. Chem. Soc. Chem. Comm.* **1983**, 1362.
98. Teo, B.K.; Shi, X.; Zhang, H. *J. Am. Chem. Soc.* **1992**, *114*, 2743.
99. (a) Schechtman, D.; Blech, I.; Gratias, D.; Cahn, J.W. *Phys. Rev. Lett.* **1984**, *53*, 1951; (b) Levine,

D.; Steinhardt, P.J. *Phys. Rev. Lett.* **1984**, *53*, 2477.

100.   Bagley, B. *Nature* **1965**, *208*, 674.
101.   Smith, D. J.; Mark, L.D. *J.Cryst. Growth* **1981**, *54*, 433.
102.   Machizaud, F.; Kuhnast, F. A.; Flechon, J. *Ann. Chim. (Paris)* **1978**, *3*, 177.
103.   (a) Smith, H. I.; Craighead, H. G. *Phys. Today* **1990**, *43(2)*, 24;   (b) Randall, J.N.; Reed, M.A.; Frazier, G. A. *J. Vac. Sci. B*, **1989**, *7(6)*, 1398.
104.   (a) Stucky, G. D.; Herron, N.; Wang, Y.; Eddy, M.; Cox, D.; Moller, K.; Bein, T. *J. Am. Chem. Soc.* **1989**, *111*, 530;   (b) Stucky, G. D.; MacDougall, J.E.; Eckert, H.; Herron, N.; Wang, Y.; Moller, K.; Bein, T. *J. Am. Chem. Soc.* **1989**, *111*, 8006.
105.   (a) Lehn, J.-M. *Angew. Chem. Int. Ed. Engl.* **1990**, *29*, 1304;   (b) Lehn, J.-M. *Pure Appl. Chem.* **1978**, *50*, 871;   (c) Lehn, J.-M. *Angew. Chem. Int. Ed. Engl.* **1988**, *27*, 89.
106.   (a) Anelli, P. L.; Spencer, N.; Stoddart, J. F. *J. Am. Chem. Soc.* **1991**, *113*, 5131;   (b) Kohnke, F. H.; Mathias, J. P.; Stoddart, J. F. *Angew. Chem. Int. Ed. Engl. Adv. Mater.* **1989**, *28*, 1103.
107.   Lindsey, J. S. *New J. Chem*, **1991**, *15*, 153.
108.   Byrn, M.P.; Curtis, C. J.; Goldberg, I.; Hsiou, Y.; Khan, S. I.; Sawin, P. A.; Tendick, S. K.; Strouse, C. E. *J. Am. Chem. Soc.* **1991**, *113*, 6549.

Hydride Ligand Location in Complexes of the Copper Triad and Other Systems

G. F. Mitchell[1], R. J. Pooley[2], A. J. Welch[1], D. A. Welch[2], and A. J. Welsh[2]

[1] DEPARTMENT OF CHEMISTRY AND [2] DEPARTMENT OF COMPUTER SCIENCE, UNIVERSITY OF EDINBURGH, UK

1 INTRODUCTION

X-ray diffraction is generally considered inadequate in locating metal-bound hydride ligands, especially for second or third row transition metals, and particularly in the cases of transition metal cluster compounds. Neutron diffraction provides the ideal solution, but such experiments are far from routine.

Accordingly a number of methods have been developed in which empirical considerations are used to suggest likely sites for hydride ligands given the non-hydrogen molecular skeleton established by an X-ray diffraction experiment. Amongst the best of these are the so-called "potential energy" methods[1,2] in which hydride ligand sites are predicted to be those in which repulsions between hydride ligands and non-hydride ligands are *minimised*. By their very nature, however, these methods fail to correctly predict hydride ligand sites in coordinatively unsaturated species, polyhydride species and interstitial hydride clusters, and for molecules such as these it is preferable to utilise a method which alternatively *maximises* attractions between the metal(s) and hydride ligand(s) present. Methods based on molecular orbital (MO) approaches are therefore appropriate.

The H ligand position in $HFe(CO)_4Mo(CO)_5$ has been successfully optimised by *ab initio* MO calculations[3] and working at this level of calculation on real molecules is clearly the ultimate objective. In the medium term, however, hydride ligand prediction *via* modified extended Hückel MO (MEHMO) calculations may be useful. EHMO calculations, which are very quick and easy to run, are generally good for probing angular interactions between atoms or fragments, but very poor at estimating optimum radial distances as no (electrostatic) repulsion terms are included. Thus in MEHMO calculations a repulsive term $WB(r)$ is added to the EHMO-calculated energy E_{EHMO} to give an overall potential $W*(r)$. Thus for a polyatomic species:

$$W*(r) = \sum W_B(r) + W_{EHMO}(r)$$

$W_B(r)$ is a measure of the interaction of nucleus A (charge Z_A) with "free atom" B, and depends on r, Z_A, and the number and Slater exponent (ζ) of the valence electrons of B, conventionally taken as the more electronegative atom. In metal hydrides H is B, and $W_H(r)$ is calculated as:

$$W_H(r) = e^{-2\zeta r}(\zeta + r^{-1})$$

2   IMPLEMENTATION

A MEHMO program (the Locator) has been written which incorporates evaluation of $W_H(r)$ in the established EHMO program ICON8. Input is the molecular skeleton established by an X-ray diffraction study (with appropriate ligand simplification) and the number of hydride ligands to be optimised. An initial routine uses this information to estimate initial position(s) for the hydride ligand(s), which are then optimised by the numerical analysis method of simplex minimisation. Thus far the program is at the stage of being tested against a number of established metal hydride topologies, but ultimately could be used predictively.

The program has an user-friendly graphical interface and is controlled by buttons and pull-down menus. A ball-and-stick representation of the molecule is displayed at all stages. Whenever a more energetically stable position for the hydride ligand is encountered the display is updated, allowing the user to monitor the progress of the experiment.

3   RESULTS

The following figures show the potential of the MEHMO method in finding hydride ligand positions. Each left hand diagram shows the molecular geometry including hydride ligand as established by an alternative method, usually a neutron diffraction study. On the right is plotted the same molecular skeleton together with the hydride ligand(s) as optimised by the program. The figures are annotated with M–H bond lengths in Å, and "difference errors" (the distance from the original neutron or X-ray position to that predicted by the Locator), also in Å.

4   CONCLUSIONS

Agreement is generally good, and in particular it is notable that the MEHMO method works well for a number of types of metal hydrides (polyhydride, co-ordinatively unsaturated, interstitial) for which the potential energy method fails. Furthermore, asymmetry in edge-bridging hydride ligands is often correctly predicted.

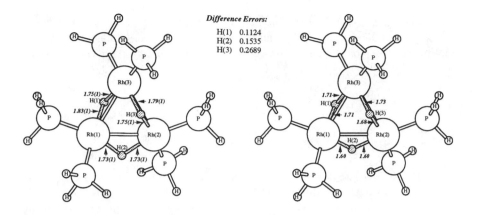

Example 1: $Rh_3H_3\{P(OCH_3)_3\}_6$, [4] *modelled by* $Rh_3H_3(PH_3)_6$, *is a coordinatively unsaturated metal hydride.*

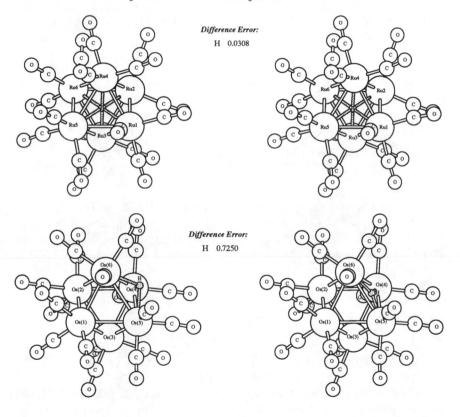

Examples 2 and 3: $[Ru_6H(CO)_{18}]^-$ [5-7] *and* $[Os_6H(CO)_{18}]^-$ [8,9] *are chemical analogues, but have topologically different hydride ligands — interstitial and face-capping respectively.*

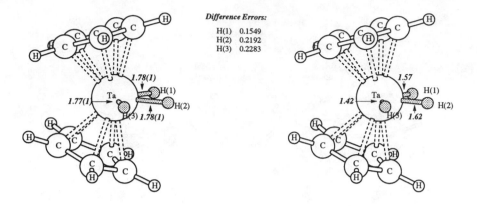

Difference Errors:

H(1) 0.1549
H(2) 0.2192
H(3) 0.2283

Example 4: $TaH_3(C_5H_5)_2$ [10] *is a polyhydride.*

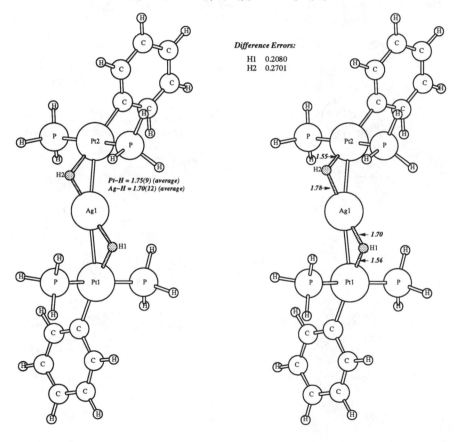

Difference Errors:

H1 0.2080
H2 0.2701

Pt–H = 1.75(9) (average)
Ag–H = 1.70(12) (average)

Example 5: $[\{(Ph_3P)_2(C_6H_5)PtH\}_2Ag]^+$,[11] *modelled by* $[\{(PH_3)_2(C_6H_5)Pt$-$H\}_2Ag]^+$, *is an example of a compound with a group 11 metal atom involved in two hydride bridges.*

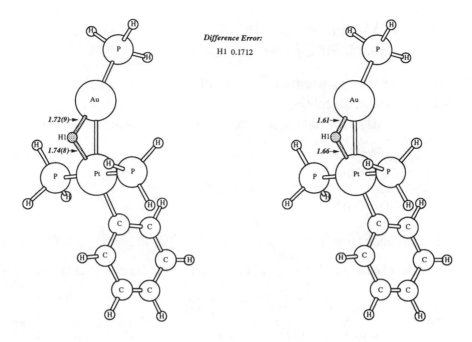

Example 6: *In $[(Ph_3P)Au(\mu_2\text{-}H)Pt(PEt_3)_2(C_6H_5)]^+$,[12] modelled by $[(PH_3)AuHPt(PH_3)_2(C_6H_5)]^+$, both metal atoms are coordinatively unsaturated.*

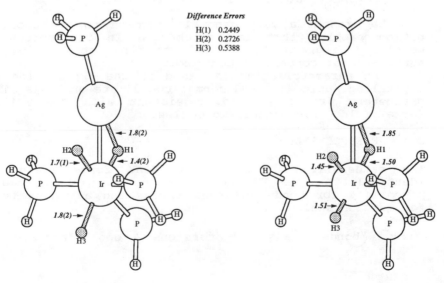

Example 7: $[(PH_3)AgH_3Ir(PH_3)_3]^+$,[13] *has two terminal and one bridging hydride ligand, all correctly located. The asymmetry in the $Ag\text{-}\mu H\text{-}Ir$ system is correctly reproduced.*

A New Algorithm and Parameters for Force Field Calculations of Copper(II) Complexes

S. Teipel[1], F. Wiesemann[1], B. Krebs[1], U. Höweler[2], and K. Angermund[3]

[1] UNIVERSITÄT MÜNSTER, ANORGANISCH-CHEMISCHES INSTITUT, D-4400 MÜNSTER, GERMANY
[2] UNIVERSITÄT MÜNSTER, ORGANISCH-CHEMISCHES INSTITUT, D-4400 MÜNSTER, GERMANY
[3] MAX-PLANCK-INSTITUT FÜR KOHLENFORSCHUNG, D-4330 MÜLHEIM/ RUHR, GERMANY

1 INTRODUCTION

Force field methods have been applied successfully to the modelling of structures of organic molecules and proteins. A variety of software packages make these methods an everyday tool for chemists.
The extension of the force fields to inorganic molecules, metal complexes and metalloproteins faces two serious problems:
- the more complex coordination spheres require new algorithms
- the high structural flexibility makes the parametrisation more difficult.

We present a very general approach to the coordination problem that is implemented in the molecular modelling programme MOBY [1] and its first application to square-planar copper(II) complexes.
The parametrisation is based on the exploration of crystallographic data and normal coordinate analyses. The parameters are determined consistent with the TRIPOS force field [2] for organic molecules.

2.1 THE FORCE FIELD

The mathematical form of the MOBY force field is identical to the TRIPOS force field with harmonic potentials for bonds and bond angles and a cosine term for torsion angles:

$$V_{tot} = V_{bonds} + V_{angles} + V_{torsions} + V_{oop} + V_{nonvalence}$$

The only difference between the MOBY and the TRIPOS force field is found for the out-of-plane potential (V_{oop}) of planar coordinated (sp^2 hybridized) atoms A:

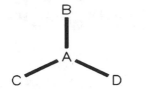

TRIPOS [2]:

$V_{oop}=k_{oop}\ d^2$;
d: distance of A to the plane BCD.

MOBY [1]:

$V_{oop}=V_{oop}(1+\cos(2\phi-180°))$;
ϕ:dihedral angle DCAB

We determined V_{oop} by a fit to the quadratic potential about the minimum.

No charges have been assigned to the atoms. Thus the non valence term consists of a Lennard-Jones 6-12 potential only.

The metal-ligand interactions are treated by valence force field terms.

2.2 THE NEW ALGORITHM

The multiple-reference value problem makes standard force field implementations inappropiate for the calculation of complex coordination polyhedra. For the tetrahedral coordination in molecules like CCl_4 only one reference value (109.5°) is needed to describe the bond angle terms. Also for the other coordination polyhedra usually found in organic compounds - trigonal planar and linear - only one reference value (120° for the planar, 180° for the linear structures) is needed.

For a square planar coordination like in $Cu(NH_3)_4^{2+}$ this approach will lead to incorrect structures. The bond angles are 90° and 180°, respectively. Recently a continuous trigonometric potential has been proposed for those situations[3].

We have introduced a very simple approach:
More than one reference value θ_0 is provided for bond angles. Each reference value is accompanied by its own force constant k_θ. The programme assigns the reference value to a bond angle potential that is closest to the actual angle in the starting structure. The user has to provide the list of reference values and can thus decide which coordination sphere should be investigated.

With this approach we are not only able to calculate coordination polyhedra like square planar, trigonal bipyramidal or octahedral, it is also possible to describe cyclic π-ligands like the cyclo-pentadienyl (CPD) anion in a consistent manner.

The metal-ligand bond is calculated like a σ-bond from the metal atom to a dummy centre in the centre of the aromatic ring. The dummy centre is bonded to the real carbon centres of the molecule. Here one needs two different reference values (72° and 144°) for the C-dummy-C angles. With this approach CPD-complexes can be optimized with satisfactory results [4].

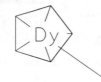

The figure shows the new model for a
π-ligand coordinated to a metal atom.

2.3 THE PARAMETRISATION

29 Cu(II) compounds with N and O ligands have been selected from the Cambridge Crystallographic Data File. After addition of hydrogen atoms these structures have been optimized to rms gradients below 0.1 kcal/mol Å. Normal coordinate analyses have been performed to check the convergence of the optimization and the force constants for the bonds and bond angles potentials.

The reference values for the copper-ligand bonds were taken from statistics on the data file. Some refinements were necessary in the course of the parametrisation. The initial force constants for the bonds were calculated from IR spectral data on sample complexes in accordance with the force constants for carbon-hydrogen bonds taken from the TRIPOS force field. The force constants for the bond angle potentials at the metal atom were obtained by comparison to available parameters.

3 RESULTS

Two sets of optimisations were run:
- starting from the crystal structure
- starting from a structure that was obtained by a random distortion of the Cartesian coordinates.

The rms values for atom positions and internal geometric parameters shows the same agreement of optimized structures to crystal geometries that has been obtained with the TRIPOS force field for organic molecules.

Table 1: Comparison of optimized structures with X-ray structures

rms deviations of internal coordinates (mean values of 29 structures)

	about all atoms	only terms with Cu	terms with heavy atoms (except Cu)	terms with hydrogen
bonds [Å]	0.07	0.03	0.03	0.13
angles [°]	2.87	1.51	0.44	3.36
torsions[°]	6.99	8.36	4.53	8.48

The differences between measured IR frequencies and the frequencies calculated by a normal coordinate analysis are less than 30 cm^{-1} for every vibration involving the Cu(II) atom.

Thus, the algorithm and the parametrisation lead to a very reliable description of square-planar coordinated Cu(II) complexes with a great variety of O and N ligands [5].

Lengths (Å) of the bonds around the Cu atom
black: calculated structure,
white: X-ray structure

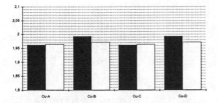

Angles (cis-angles=φ-90°, trans-angles=φ-180°) at the Cu atom
black: calculated structure,
white: X-ray structure

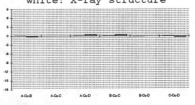

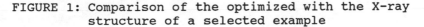

FIGURE 1: Comparison of the optimized with the X-ray structure of a selected example

REFERENCES

1   U. Höweler, 'MOBY, Molecular Modelling on the PC', Springer-Verlag, Berlin, 1991, Version 1.5.

2   M. Clark, R.D. Cramer and N.V. Opdenbosch, J. Comp. Chem., 1989, 10, 982.

3   V.S. Allured, C.S. Kelly and C.R. Landis, J. Am. Chem. Soc., 1991, 113, 1.

4   M. Nolte, Dissertation, University of Münster, Münster, 1992

5   F. Wiesemann, S. Teipel, B. Krebs, U. Höweler and K. Angermund, J. Comp. Chem., to be submitted.

Structures and Magnetic Properties of μ_4-Oxo Bridged Copper and Homodinuclear Zinc Complexes

S. Uhlenbrock[1], S. Teipel[1], B. Krebs[1], K. Griesar[2], and W. Haase[2]

[1] UNIVERSITÄT MÜNSTER, ANORGANISCH-CHEMISCHES INSTITUT, D-4400 MÜNSTER, GERMANY

[2] TH DARMSTADT, PHYSIKALISCH-CHEMISCHES INSTITUT, D-6100 DARMSTADT, GERMANY

1 INTRODUCTION

Since Bertrand et al. [1] characterized the first tetranuclear copper(II) complex of the general formula $[Cu_4OX_{10-n}L_n]^{n-4}$ where X represents a halide ion and L represents a Lewis base ligand in 1966, a number of these compounds with the special formula $[Cu_4OX_6L_4]$ (X = Br$^-$, Cl$^-$) has been synthesized. We here report on synthesis and structures of two novel Cu^{II} complexes of a new type which also show this special feature of a central μ_4-oxo bridge.

Phospholipase C from Bacillus cereus is a zinc enzyme which catalyzes the hydrolytic cleavage of the bond between the phosphate ester and the glycerol moiety in phospholipids. As a result of the X-ray structure analysis there are three zinc atoms in the active site [2]. Zn(1) and Zn(3) are connected by one OH$^-$ or H_2O bridge and one carboxylato group. The Zn···Zn distance is 3.3 Å. All three metal atoms are in trigonal bipyramidal coordination.

2 RESULTS AND DISCUSSION

a) Copper

We have employed a new binucleating ligand, 2,6-bis(morpholinylmethyl)-4-methylphenol (Hbmmk), obtained by a Mannich reaction between a para-substituted phenol, formaldehyde, and morpholine [3]. With this tridentate ligand two tetramers of the new type $[Cu_4OX_4L_6]$ could be prepared by reaction of Hbmmk with $CuBr_2$ or $Cu(ClO_4)_2 \cdot 6H_2O$ and NaOBz in methanol [4]:

$[Cu_4OBr_4(bmmk)_2] \cdot 2MeOH$ <u>1</u>

$[Cu_4O(OBz)_4(bmmk)_2] \cdot H_2O$ <u>2</u>

The structure of the coordination sphere of <u>1</u> is shown in figure 1.

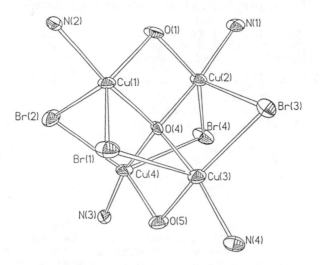

<u>Figure 1:</u>   Coordination sphere of <u>1</u>

These μ_4-oxo-bridged   cluster   compounds   are   of particular interest because one of the normally three halides in the basal plane of the trigonal bipyramidal arrangement around the copper ions is replaced by an oxygen atom in <u>1</u>. The coordination polyhedra are unusually highly distorted due to the sterical strain of the chelating ligand. <u>2</u> is the first complex of this type in which all halide ions are replaced by benzoate groups. The copper centers are no longer in trigonal bipyramidal coordination, but in two cases square pyramidal, one is square planar whereas the fourth one is in octahedral coordination. Nevertheless, a nearly perfect tetrahedron of copper(II) is observed around the central oxygen in <u>1</u>, and a slightly distorted tetrahedron in <u>2</u>.

In addition to the X-ray molecular structures, the magnetic and spectroscopic properties were studied in detail. To describe the magnetic behaviour of the compounds the tetrameres can be devided into a Cu_2O_2-system with the coupling constant J_{12} and a Cu_2O-system with the coupling constant J_{13}. The results of the calculations are collected in table 1.

<u>Table 1:</u> Magnetic properties of the complexes

complex	g	J_{12} [cm^{-1}]	J_{13} [cm^{-1}]	x_p [%]
<u>1</u>	2.19	−295	−43	2.4
<u>2</u>	2.01	−175	−24	0.8

The electronic absorption spectra of the complexes show a band at 292 nm ($\underline{1}$, ε = 11100 M^{-1} cm^{-1}; $\underline{2}$, ε = 14800 M^{-1} cm^{-1}), due to morpholine-N-to-copper(II) charge-transfer transitions. Complex $\underline{1}$ exhibits a phenolate-to-Cu(II) CT band at 337 nm (ε = 3550 M^{-1} cm^{-1}) and a broad band at 730 nm (ε = 112 M^{-1} cm^{-1}), assigned as the sum of the d-d transitions. These transitions shift to 342 nm (ε = 7030 M^{-1} cm^{-1}) and 675 nm (ε = 298 M^{-1} cm^{-1}) for complex $\underline{2}$. The large shift in the position of the d-d transition to higher energies as compared to the one of complex $\underline{1}$ results from an increased overlap of the bridging phenolate p orbitals with the copper $d_{x^2-y^2}$ orbitals in $\underline{2}$. This results from the different coordination geometries in $\underline{2}$ which are essentially square planar with additionally bonded benzoate oxygen atoms in apical positions.

b) Zinc

The aim of our present work is to model the dinuclear part of the active site of PLC. Therefore we used the heptadentate ligand N,N,N',N'-tetrakis(2-benzimidazolylmethyl)-2-hydroxy-1,3-diamino-propane (Htbpo) to stabilize homodinuclear zinc complexes.

Reaction of Htbpo with $Zn(ClO_4)_2 \cdot 6H_2O$ in isopropanol and addition of NaOAc in methanol yields

$$[Zn_2(tbpo)(OAc)](ClO_4)_2 \cdot MeOH \cdot 3H_2O \qquad\qquad \underline{3}$$

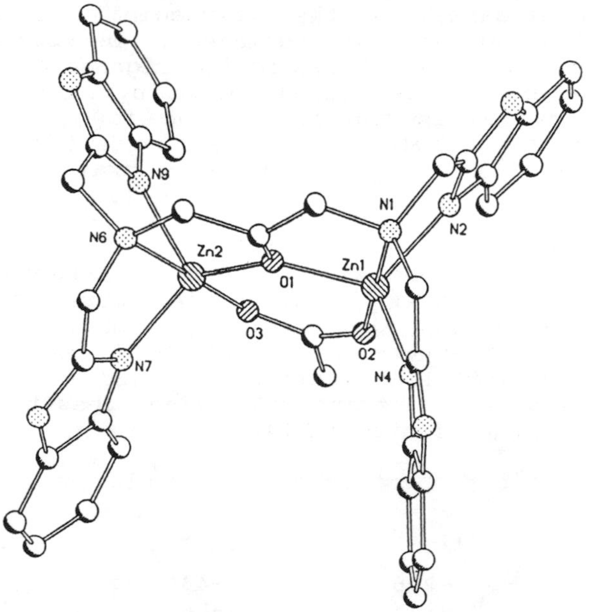

Figure 2:   Structure of $[Zn_2(tbpo)(OAc)]^{2+}$

which could be recrystallized from a methanol/ethanol mixture. The X-ray structure analysis of 3 shows a homodinulcear zinc complex which is represented in figure 2.

The molecule consists of two Zn atoms, linked by the oxygen atom of the ligand and an acetate bridge. The N-donor atoms of tbpo⁻ are completing the coordination at the metal atoms. Each zinc is surrounded by five atoms forming a slightly distorted trigonal bipyramid. The Zn···Zn distance is 3.440(3) Å.

The reaction of tbpo⁻ with $Zn(ClO_4)_2 \cdot 6H_2O$ and pyrazole in isopropanol yields a white powder of 4 which could be recrystallized from a methanol/ethanol mixture.

$$[Zn_2(tbpo)(pyrazolate)](ClO_4)_2 \cdot 4MeOH \cdot EtOH \quad \underline{4}$$

The structure of 4 is similar to 3. Again the donor atoms from the ligand and from the pyrazolate bridge are coordinating each metal ion in form of a trigonal bipyramid. Like in 3 the bond distance between the aliphatic nitrogen atoms, which are in the apical position of the coordination polyedron, and the zinc atoms are significantly longer (average 2.389 Å) than all other coordination bonds (average 1.991 Å). The distance between the two metal ions is 3.379(2) Å.

Both homodinuclear zinc complexes can be seen as a structural model for PLC. Like in the enzyme each zinc ion is coordinated to form a trigonal bipyramid. Especially 3 models very exactly the chemical surrounding of the dinuclear part of the enzyme with the alkoxo atom and the carboxylate group bridging the two metal atoms. This is also shown in the Zn···Zn distance which is very similar to the one in the enzyme.

REFERENCES

[1]   J. A. Bertrand and J. A. Kelley,
      J. Am. Chem Soc., 1966, 88, 4746.
[2]   E. Hough, L. K. Hanse, B. Birkness, K. Jynge,
      S. Hansen, A. Hordvik, C. Little, E. Dodson
      and Z. Derewenda, Nature, 1989, 338, 357.
[3]   B. Bremer, K. Schepers, S. Teipel and B. Krebs
      J. Inorg. Biochem., 1991, 43, 544.
[4]   S. Teipel, B. Krebs, K. Griesar and W. Haase
      Inorg. Chem., to be submitted.

Unusual Stabilization of Anion Radicals by Copper(I)

W. Kaim[1], M. Moscherosch[1], S. Kohlmann[1], J. S. Field[2], and D. Fenske[3]

[1] INSTITUT FÜR ANORGANISCHE CHEMIE DER UNIVERSITÄT, PFAFFENWALDRING 55, D-7000 STUTTGART 80, GERMANY

[2] UNIT OF METAL CLUSTER CHEMISTRY, DEPARTMENT OF CHEMISTRY, UNIVERSITY OF NATAL, PO BOX 375, PIETERMARITZBURG 3200, SOUTH AFRICA

[3] INSTITUT FÜR ANORGANISCHE CHEMIE DER UNIVERSITÄT, POSTFACH 6380, D-7500 KARLSRUHE, GERMANY

1 INTRODUCTION

Anion radicals, i.e. organic molecules with one additional electron are frequently prepared by reduction with alkali metals to form "ion pairs" with the resulting monovalent metal ions. Although EPR spectroscopic and crystallographic information is available on some of these ion pairs,[1] the high coordination numbers and the weak bonds between alkali metal ions and ligand atoms render these complexes quite labile.

Among the other stable monovalent metal ions available Cu^+ has found little use in coordination with anion radicals.[2] This is surprising since Cu(I), if stabilized by acceptor ligands, is quite resistant towards reduction to elemental copper; Cu(I) typically forms complexes with rather low coordination numbers and a flexible coordination geometry.[3] From the spectroscopic point of view, Cu(I) as a closed-shell d^{10} centre permits the EPR detection of just slightly perturbed radical ligands.[2,4,5] The two isotopes ^{63}Cu (69.2%) and ^{65}Cu (30.8%) both have a nuclear spin of I = 3/2 and relatively large nuclear magnetic moments which differ by about 7%.[5]

In this contribution we show that cationic Cu(I) complex fragments $^+Cu(PR_3)_2$ can stabilize bis-chelating anion radicals to such an extent that crystalline paramagnetic dicopper(I) cations can be formed under atmospheric conditions in reactions involving elemental copper as reductant.

2 REACTIVITY AND ELECTROCHEMISTRY

The three cationic complexes which are discussed in this contribution are illustrated below (1); all have been isolated and characterised, complexes **1** and **2** by means of X-ray crystallography.

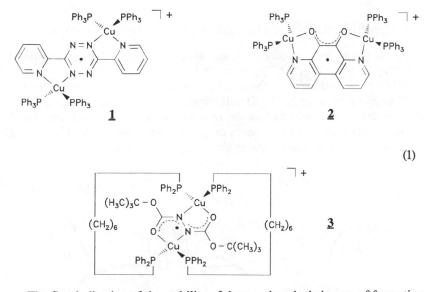

(1)

The first indication of the stability of these cations is their ease of formation. Complex **1** with the 3,6-bis(2-pyridyl)-1,2,4,5-tetrazine anion radical as the bridging bis-chelate ligand was obtained from the diamagnetic dication (E_{red} = +0.13 V vs. SCE)[4] by reduction with zinc in CH_3OH/CH_2Cl_2. Compound **2** with N,O;N',O'-chelating 4,7-phenanthroline-5,6-semidione as the bridging ligand was obtained simply by reacting the non-reduced dione with excess $[Cu(PPh_3)_4](BF_4)$ in dichloromethane.[4] Cyclic voltammetry showed that neither oxidation nor reduction of this paramagnetic cation is reversible[4] - the radical intermediate is the only stable oxidation state of this quinonoid redox system! The dinuclear complex **3** undergoes electro-chemically reversible oxidation and reduction,[6] however, the two potentials are separated by 1.16 V, corresponding to a comproportionation constant of $10^{19.7}$ of the paramagnetic intermediate. Deep blue compounds of type **3** were formed by reacting di-tert.-butyl-azodicarboxylate, arylphosphine and elemental copper in wet methanol under air.[5,6] This unusual reaction may proceed via electron transfer catalysis involving O_2/O_2^- and $Cu/Cu^+/Cu^{2+}$;[5] but, in any case, it illustrates the high tendency of formation for this type of paramagnetic compounds.

3 EPR SPECTROSCOPY

The claim that complexes **1-3** contain Cu(I) centers bridged by an anion radical ligand is substantiated by EPR spectroscopy which shows the expected ligand hyperfine structures, high resolution and low g anisotropy.[4,5] There is no evidence for alternative formulations which would involve a dianionic bis-chelate ligand and a mixed-valent Cu(I)/Cu(II) situation or a neutral bridge with Cu(0)/Cu(I) centres. Nevertheless, the ^{31}P, ^{63}Cu and ^{65}Cu coupling constants can be conveniently determined[5] and, indeed, the molecular structures show the possibility for overlap between Cu-P bond orbitals and the π system of the bridging radical ligand.

4 STRUCTURE

Crystal structures were obtained for the tetrafluoroborate salt of **1** and for the tetraphenylborate salt of **3**.[6] The structure of the cation **1** is as expected with "normal"[7] tetracoordination at the Cu(I) atoms and coplanar six-membered rings of the bridging radical (Fig. 1). The theoretical and spectroscopic result that the unpaired electron is localized on the central tetrazine ring[4] is confirmed structurally, e.g., the corresponding bond distances in **1** lie between those of tetrazine $C_2H_2N_4$ and 2e-reduced 1,4-dihydrotetrazine.[8]

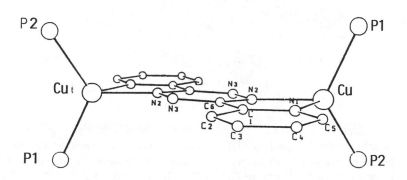

<u>Figure 1</u> Structure of the cation **1** in the crystal. Phenyl groups of PPh_3 ligands and hydrogen atoms are omitted for clarity.

The complex cation **3** has the two approximately tetrahedrally coordinated copper centers bridged by a virtually planar 6 centre π system of the azodicarbonyl anion radical ligand. Remarkably, the long chain diphosphines do not chelate at a single metal centre to give energetically unfavourable nine-membered rings; instead, the rather small distance of 4.82 Å imposed by the geometry of the edge-shared five-membered chelate rings of the S-frame bridging ligand radical allows additional intramolecular bridging of the two copper(I) centres. The metals are thus bridged on the outside by two diphosphines and from within by the bis-chelating radical ligand (Fig. 2). The eight phenyl rings and two tert.-butyl groups are positioned so as to effectively protect the metal/radical/metal core. Complex **3** may be viewed[6] as an "inverse cryptate" in which a central radical anion is stabilized through coordination by two copper(I) centers and protected from the environment. The short Cu-N (2.02 Å) but long Cu-O bonds (2.34 Å) suggest that coordination of the soft Cu(I) centres favours a hydrazido(2-) formulation which is also the result of a $2e^-/2H^+$ reduction of azodicarboxylates.

Several factors may favour the remarkable stabilization of anion radical intermediates by $^+Cu(PR_3)_2$ complex cations. Among these is the positive charge, the small coordination number and structural flexibility of Cu(I), the stability of the Cu(I)-P bond, a certain amount of ligand-to-metal delocalization of charge and spin, and the steric protection by arylphosphine ligands.

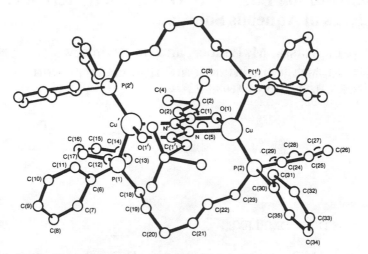

<u>Figure 2</u> Structure of the cation **3** in the crystal. Hydrogen atoms are omitted for clarity.

The ultimate steric shielding is provided by proteins in enzymes where a Cu(I)-semiquinone state was established recently.[9] The effect of steric protection is absent for complexes between $^+$ZnR, another positively charged d^{10} system, and chelating anion radicals which allows the dimerization of such paramagnetic species.[10] Heavier homologues of Cu(I) and Zn(II) have a rather high threshhold potential for reduction to the element in the coordinated state which clearly limits their use in coordination to one electron-reduced ligands.

REFERENCES

1. H. Bock et al., <u>Angew. Chem.</u>, 1992, <u>104</u>, 564.
2. G.A. Razuvaev, V.K. Cherkasov and G.A. Abakumov, <u>J. Organomet. Chem.</u>, 1978, <u>160</u>, 361.
3. C. Vogler, H.-D. Hausen, W. Kaim, S. Kohlmann, H.E.A. Kramer and J.Rieker, <u>Angew. Chem. Int. Ed. Engl.</u>, 1989, <u>28</u>, 1659, and literature cited.
4. W. Kaim and S. Kohlmann, <u>Inorg. Chem.</u>, 1987, <u>26</u>, 1469.
5. W. Kaim and M. Moscherosch, <u>J. Chem. Soc., Faraday Trans.</u>, 1991, <u>87</u>, 3185.
6. M. Moscherosch, J.S. Field, W. Kaim, S. Kohlmann and M. Krejcik, <u>J. Chem. Soc., Dalton Trans.</u>, submitted.
7. W. Kaim, S. Kohlmann, J. Jordanov and D. Fenske, <u>Z. Anorg. Allg. Chem.</u>, 1991, <u>598/599</u>, 217.
8. S. Kohlmann, <u>Ph.D. Thesis</u>, University of Frankfurt/Main, 1988.
9. D.M. Dooley, M.A. McGuirl, D.E. Brown, P.N. Turowski, W.S. McIntire and P.F. Knowles, <u>Nature</u>, 1991, <u>349</u>, 262.
10. M. Kaupp, H. Stoll, H. Preuss, W. Kaim, T. Stahl, G. van Koten, E. Wissing, W.J.J. Smeets and A.L. Spek, <u>J. Am. Chem. Soc.</u>, 1991, <u>113</u>, 5606.

The Substitution Lability of Gold(I) and Mercury(II) Complexes in Aqueous Solution

G. Geier, F. A. Issa, M. Lauber, and M. Szvircsev

LABORATORIUM FÜR ANORGANISCHE CHEMIE, ETH ZÜRICH,
CH-8092 ZÜRICH, SWITZERLAND

1 GOLD(I) COMPLEXES

Several gold(I) complexes are reduced by H_2O or they are subject to disproportionation in aqueous solution. However, a number of soft ligands stabilize the oxidation state $+I$, e.g. CN^-, thiolates, and thiones. These ligands form very stable linear complexes, with a wide range of stabilities i.e., between $\log\beta_2 = 39$ for CN^- ($\beta_2 = [Au(CN)_2^-]/[Au^+][CN^-]^2$) and $\log\beta_2 = 23.3$ for 1-methylpyridine-2-thione (MPT).[1] We have determined the equilibrium and rate constants for the substitution reactions for several ligands L and X (1). (Charges have been omitted in (1)). If L or X are basic ligands, equilibria (1)

$$AuL_2 + 2X \rightleftharpoons AuLX + L + X \rightleftharpoons AuX_2 + 2L \qquad (1)$$

can be shifted by protonation of the ligands, thus allowing the study of the substitution reactions for ligands of very different coordination tendencies.

Equilibria and Kinetics with S-Donor Ligands

In particular, we have investigated the equilibria and kinetics of substitution reactions (1) for X = 4-nitro-2-sulphonato-benzenethiolate ($NTPS^{2-}$) and L = thiourea (TU) and MPT. $NTPS^{2-}$ and MPT are chromophoric ligands. Their UV-vis spectra change on coordination and protonation (pK_a of $H(NTPS)^- =$

5.20).[1,2] Thus the equilibrium constants for the first step,
$K_1 = [AuL(NTPS)] [L]/[AuL_2][NTPS]$, and second step,
$K_2 = [Au(NTPS)_2][L]/[AuL(NTPS)] [NTPS]$,
were determined by spectrophotometry, cf. Table 1.

Stepwise substitution equilibria (1) are typically overlapping, and they are rapidly established. This makes it impossible to obtain samples containing only the mixed-ligand complexes $AuL(NTPS)^-$ in solution. This is also reflected by the ligand-redistribution equilibrium (2) with the equilibrium constant,

$$AuL_2^+ + Au(NTPS)_2^{3-} \rightleftharpoons 2Au(NTPS)L^- \tag{2}$$

$K_{rd} = 16$ for L = TU and MPT. In spite of this difficulty, it was possible to study the kinetics of the stepwise substitutions as quasi-onestep reactions with the stopped-flow method.

<u>Kinetics of the First Step.</u>   The kinetics of step 1 were studied under pseudo-first-order reversible conditions with $[AuL_2^+]$ and [L] in excess, and pH << $pK_a = 5.20$. The rate expression for the mechanism according to

$$k_{obs} = (k_1 K_a/[H^+] + k_{H1})[AuL_2^+] + (k_{-1} + k_{-H1})[L] \tag{3}$$

Scheme 1 is expressed by equation (3). The rate constants were obtained from the various concentration and pH dependencies (pH 1.0 - 2.8), cf. Table 1.

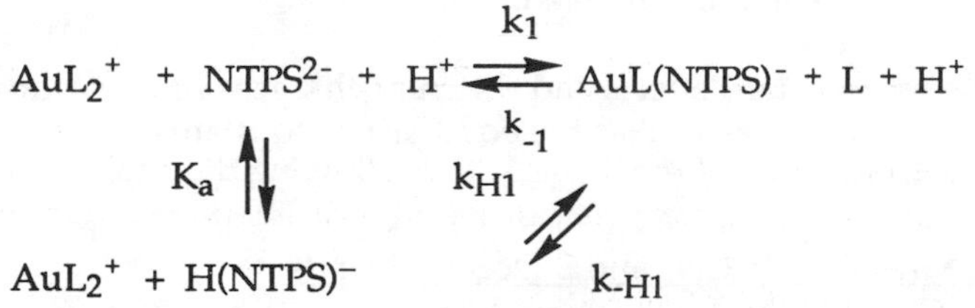

$$AuL_2^+ + NTPS^{2-} + H^+ \underset{k_{-1}}{\overset{k_1}{\rightleftharpoons}} AuL(NTPS)^- + L + H^+$$

$$K_a \updownarrow \qquad \qquad k_{H1} \nearrow\nearrow$$

$$AuL_2^+ + H(NTPS)^- \qquad \qquad k_{-H1}$$

<u>Scheme 1</u>

The non-basic S-donor ligands TU and MPT exhibit similar reactivity as complexing agents for Au(I). The rates of substitution by $NTPS^{2-}$ approach almost the diffusion-controlled limit. This may be surprising if one takes into account that the bond between Au(I) and S(-II) is typically covalent. $H(NTPS)^-$ is at least 4 - 5 orders of magnitude less reactive than $NTPS^{2-}$, cf. k_1/k_{H1}.

Table 1   Rate[a] and equilibrium constants for substitution reactions
               in the Au(I)/NTPS^{2-}/L-system, 25 °C

Step 1

L	k_1	k_{-1}	k_{H1}	$\log(k_1/k_{-1})$	$\log K_1$
MPT	1.9×10^9	2.3×10^5	$\leq 3 \times 10^4$	3.9	3.8
TU	1.6×10^9	2.6×10^5		3.8	3.9

Step 2

L	k_2	k_{-2}		$\log(k_2/k_{-2})$	$\log K_2$
MPT	5.6×10^7	1.7×10^5		2.5	2.6
TU	1.7×10^7	7.5×10^4		2.4	2.7

[a] k in M^{-1}s^{-1}.

Kinetics of the Second Step.   Similarly to the first step, we started with
solutions of AuL$_2^+$. However, pseudo-first-order reversible conditions were
established by using an excess of [NTPS]$_t$ and [L]$_t$. In this way the first step is
completed already within the mixing time. The accessible pH range is limited,
because at low pH values the first step interferes with the second, and above
pH 3 the rates are too fast for the stopped-flow method. A reaction mechanism
analogous to Scheme 1 could be obtained.

Allowing for statistical effects and different charges of the reacting
species, the rate constants, as well as the equilibrium constants, are similar to
those of the first step.

Equilibria and Kinetics with CN$^-$ and S-Donor Ligands

An analogous study of the reaction system (1) was carried out for X =
CN$^-$ (pK$_a$ = 9.01) and L = TU, MPT, and NTPS^{2-}. The rate and equilibrium
constants for step 2 (see eq. 4) are given in Table 2. The rate constants k$_2$ show

$$AuL(CN) + CN^- \underset{k_{-2}}{\overset{k_2}{\rightleftharpoons}} Au(CN)_2^- + L \tag{4}$$

that, besides the charge effects, there is also a correlation between rate and
equilibrium constants. This is in line with an associative (I$_a$) mechanism. By
comparing the rate constants in Table 2 with those in Table 1 it seems likely
that the intrinsic rates for substitutions of S-donor ligands by CN$^-$ are slightly
slower than those for substitutions of S-donor ligands by NTPS^{2-}.

Table 2   Rate and equilibrium constants for substitution reactions
in the Au(I)/CN⁻/L-system, 25 °C

L	$k_2(M^{-1}s^{-1})$	$k_{-2}(M^{-1}s^{-1})$	$logK_2$
TU	2.8×10^8	4.0	7.8
MPT	1.9×10^8	1.4	8.1
NTPS²⁻	7.6×10^6	1.9×10^2	4.6

2   MERCURY(II) COMPLEXES

The substitution lability of the isoelectronic d^{10} center Hg(II) is even more
pronounced.[2,3] The similar substitution reactions (5) and (6) show clearly

$$Au(TU)_2^+ \; + \; NTPS^{2-} \; \underset{k_{-1}}{\overset{k_1}{\rightleftharpoons}} \; Au(TU)(NTPS)^- \; + \; TU \qquad (5)$$

$$CH_3Hg(TU)^+ \; + \; NTPS^{2-} \; \underset{k_{-1}}{\overset{k_1}{\rightleftharpoons}} \; CH_3Hg(NTPS)^- \; + \; TU \qquad (6)$$

that with $CH_3Hg(II)$ the diffusion-controlled limit is reached, whereeras the
statistically corrected value for Au(I) is almost one order of magnitude lower,
cf. Table 3. Generally, substitution reactions of $CH_3Hg(II)$ complexes with
various types of ligand are diffusion controlled.[2]

Table 3   Comparison between the substitution lability
of $CH_3Hg(II)$ and Au(I) according to (5) and (6)

	$k_1(M^{-1}s^{-1})$	$k_{-1}(M^{-1}s^{-1})$	$logK$
Au(I)	1.6×10^9	2.6×10^5	3.8
statist. corr.	8×10^8		
$CH_3Hg(II)$	5×10^9	2.0×10^5	4.4

REFERENCES

1.   P. N. Dickson, A. Wehrli, and G. Geier, Inorg. Chem., 1988, 27, 2921.
2.   G. Geier, and H. Gross, Inorg. Chim. Acta, 1989, 156, 91.
3.   H. Gross and G. Geier, Inorg. Chem., 1987, 26, 3044.

Subject Index